Image Atlas of COVID-19

Image Atlas of COVID-19

MEIYUN WANG
Department of Medical Imaging, Henan Provincial People's Hospital, Zhengzhou, Henan, P. R. China

GEORGES EL FAKHRI
Gordon Center for Medical Imaging, Department of Radiology, Harvard Medical School (HMS), Boston, MA, United States

ELSEVIER

ACADEMIC PRESS
An imprint of Elsevier

Academic Press is an imprint of Elsevier
125 London Wall, London EC2Y 5AS, United Kingdom
525 B Street, Suite 1650, San Diego, CA 92101, United States
50 Hampshire Street, 5th Floor, Cambridge, MA 02139, United States
The Boulevard, Langford Lane, Kidlington, Oxford OX5 1GB, United Kingdom

B&R Book Program

Notices

Knowledge and best practice in this field are constantly changing. As new research and experience broaden our understanding, changes in research methods, professional practices, or medical treatment may become necessary.

Practitioners and researchers must always rely on their own experience and knowledge in evaluating and using any information, methods, compounds, or experiments described herein. In using such information or methods they should be mindful of their own safety and the safety of others, including parties for whom they have a professional responsibility.

To the fullest extent of the law, neither the Publisher nor the authors, contributors, or editors, assume any liability for any injury and/or damage to persons or property as a matter of products liability, negligence or otherwise, or from any use or operation of any methods, products, instructions, or ideas contained in the material herein.

ISBN: 978-0-443-18493-2

For Information on all Academic Press publications
visit our website at https://www.elsevier.com/books-and-journals

Publisher: Jonathan Simpson
Acquisitions Editor: Glyn Jones
Editorial Project Manager: Naomi Robertson
Production Project Manager: Swapna Srinivasan
Cover Designer: Christian J. Bilbow

Typeset by MPS Limited, Chennai, India

Contents

List of contributors

Yan Bai
Henan Provincial People's Hospital, Zhengzhou, Henan, P. R. China

Zehua Shao
Fuwai Central China Cardiovascular Hospital, Zhengzhou, Henan, P. R. China

Xiaochen Li
Henan Provincial People's Hospital, Zhengzhou, Henan, P. R. China

Fengmin Shao
Henan Provincial People's Hospital, Zhengzhou, Henan, P. R. China

Zhiping Guo
Fuwai Central China Cardiovascular Hospital, Zhengzhou, Henan, P. R. China

Qingxia Wu
Henan Provincial People's Hospital, Zhengzhou, Henan, P. R. China

About the authors

Dr. Meiyun Wang, MD, PhD, is a neuroradiologist, professor, and chair of the Medical Imaging Center of Henan Province, and chair of the Medical Imaging Institute and the Department of Medical Imaging of the Henan Provincial People's Hospital, associate dean of the College of Medical Technology, Zhengzhou University. She serves as a Board of Trustee member of the International Society for Magnetic Resonance in Medicine (ISMRM), certified teacher in Clinical MRI of ISMRM and the European Society for Magnetic Resonance in Medicine and Biology. She is the vice-chair of the Committee on Women in Chinese Society of Biomedical Engineering (CSBME) and a standing member of the Artificial Intelligence Committee in CSBME and Chinese Society of Radiology. Also, she is acting in her capacity as the 2020–21 president of the Overseas Chinese Society for Magnetic Resonance in Medicine (OCSMRM), chair of the Psychiatric MR Spectroscopy and Imaging Study Group of ISMRM, president-elect of the International Society for Neurovascular Disease, and vice-chair of the Chinese Chapter of the ISMRM. Not only has Dr. Wang served on the Annual Meeting Program Committee, the MR Value Initiative Committee, the Chapters Committee, and the Ad Hoc Committee on Women in ISMRM but also she spent 2 years as a postdoc in the A.A. Martinos Center for Biomedical Imaging, Massachusetts General Hospital, Harvard Medical School from 2006 to 2008. With 27 years of working experience in diagnostic MR imaging and MRI research, she focused her research efforts in exploring the value of multimodel imaging including CT, MRI, PET/CT, PET/MR, and MR-guided focused ultrasound and transcranial magnetic stimulation to improve diagnosis and treatment of diseases. Dr. Wang has published over 300 papers in journals such as *JAMA*, *Nature Medicine*, *Nature Communications*, and *Radiology*; and she has given over 300 invited lectures at international and national scientific and educational meetings. She is also the principal investigator on international cooperative grants from the National Natural Science Foundation of China and the National Key R&D Program of China. Owing to her contributions to medical field, she was awarded the American Institute for Medical and Biological Engineering (AIMBE) Fellow, ISMRM Fellow, Outstanding Contribution Award by the OCSMRM, and the Gold Medal of the Chinese Society of Radiology.

Dr. Georges El Fakhri is the Nathaniel and Diana Alpert Professor of Radiology at Harvard Medical School (HMS) and the founding director of the Gordon Center for Medical Imaging at Massachusetts General Hospital (MGH) and HMS. He is also co-director of the Division of Nuclear Medicine and Molecular Imaging and associate chair for the Medical Sciences of the Department of Radiology at MGH. Dr. El Fakhri is an internationally recognized expert in quantitative molecular imaging (SPECT-CT, PET-CT, and PET-MR) and for the in vivo assessment of pathophysiology in brain, cardiac, and oncologic diseases. His current areas of research include high-resolution PET and MR imaging in a range of diseases including neurodegenerative disease and traumatic brain injury (amyloid and neurofibrillary tangles), cardiac disease (mitochondrial membrane potential), as well as guiding radiotherapy planning (PET/MRSI). Having authored or coauthored over 300 papers and mentored over 100 students, postdocs and faculty and having delivered over 300 invited talks, keynotes, and plenary lectures, Dr. El Fakhri has received many awards and honors, including the Mark Tetalman Award from the Society of Nuclear Medicine, the Dana Foundation Brain and Immuno-Imaging Award, the Howard Hughes Medical Institutes Training Innovation Award, and the Edward J. Hoffman Award from the Society of Nuclear Medicine and Molecular Imaging. He was elected Fellow to the SNMMI, AAPM, AIMBE, IAMBE, and IEEE for "contributions to biological imaging."

Preface

The history of human development is a history of the fight against infectious diseases. In history, there have been dozens of infectious diseases that have endangered human beings, such as plague, smallpox, cholera, leprosy, diphtheria, influenza, syphilis, typhus, malaria, rabies, and tuberculosis. Among these diseases, plague, smallpox, and influenza have had the greatest impact. Since the beginning of the 21st century, SARS (severe acute respiratory syndrome), avian influenza, Ebola hemorrhagic fever, MERS (Middle East respiratory syndrome), and other epidemics have recurred frequently. In the past 2 years coronavirus disease 2019 (COVID-19) has spread to almost every country in the world, and the situation of humans fighting against infectious diseases is still severe.

As medical workers fighting on the front line of COVID-19 prevention and control, we are not only obliged to provide high-quality and efficient medical and health services but also responsible for seeking the general rules of effective intervention and optimal service supply in the context of the COVID-19 pandemic from an academic view.

Image Atlas of COVID-19 collects CT, DR, PET/CT, and PET/MR images of 61 typical cases. Each image has been interpreted in detail, and some imaging examinations were performed after patient discharge. The purpose was to support the improvement of the clinical treatment plan by compiling the imaging data of successfully treated cases.

As COVID-19 is an emerging disease, the research progress is constantly in the process of getting updated, coupled with the limited disease data and preparation time, so that there may be inadvertent omissions and deficiencies in this book. All the criticism, updates, and suggestions are warmly welcomed.

Introduction

The novel coronavirus pneumonia was named "Corona Virus Disease 2019"(COVID-19) by WHO on February 11, 2020. The International Committee on Taxonomy of Viruses has named the novel coronavirus "Severe Acute Respiratory Syndrome Coronavirus 2" (SARS-CoV-2).

COVID-19 is a novel emerging respiratory infectious disease. Although the gold standard for clinical diagnosis of COVID-19 remains the nucleic acid test for SARS-CoV-2, chest CT scans have been widely used in the diagnosis and treatment of COVID-19 as simple and fast diagnostic methods indispensable for their quick turnaround and accuracy. Chest CT scans can prove crucial when faced with a large volume of symptomatic and asymptomatic patients, and the need is for a sensitive diagnostic tool to support medical treatment. The imaging manifestations of COVID-19 have some specific characteristics, and a better understanding of the imaging findings is of great value for detecting lesions, evaluating lesion severity, diagnosis and differential diagnosis, assessment of therapeutic options, and prognosis for COVID-19 patients.

This book includes a collection of over 400 DR, CT, PET/CT, and PET/MR studies in 61 clinical cases of patients with proven COVID-19 that are detailed in this work. The book is organized into five chapters by clinical subtypes of COVID-19. The book details a large number of clinical images with illustrations and quantitative analyses intended to demonstrate imaging findings of COVID-19 to serve as a reference manual for medical staff engaged in imaging and clinical work, and as the training material for academic education.

This book and future editions are planned to include new findings as they become available from centers fighting COVID-19 around the world. The authors hope that readers will share their experiences in the imaging and treatment of COVID-19.

CHAPTER 1

Computed tomography manifestations and prognosis of mild COVID-19 cases

Section 1: Computed tomography manifestations of asymptomatic patients with familial cluster transmission

Cases 1–6

Medical history

Patient 1 is a 20-year-old woman who lives in the epidemic area and goes to visit her hometown. In the initial stage of her visit, she met with her father (patient 2) and mother (patient 6) the day she came home to them. Three days after her staying with her parents, she accompanied 5 relatives (patients 2–6, her parents included) to visit another hospitalized relative at a local hospital. It's notable that there were no reported COVID-19 cases in that hospital then. Patient 6 reported fever and sore throat 7 days after patient 1 was staying at home with them. Patients 2–5 reported fever and respiratory symptoms 13–16 days later and were admitted to the hospital. The nucleic acid tests for patients 2–6 were all positive. Patient 1 was quarantined after the disease onset of her family members, but she denied fever, respiratory, and gastrointestinal problems, and her chest Computed tomography (CT) images between 17 and 21 days after her visit back home showed no remarkable changes. The nucleic acid test for patient 1 was negative on the 16th, 26th, and 29th day and positive on the 18th day after her visit back home [1].

Laboratory findings

Patient 1: White blood cell count: 5.65×10^9/L, lymphocyte count: 1.55×10^9/L, lymphocyte percentage: 27.5%, C-reactive protein: 0.69 mg/L.

Patient 2: White blood cell count: 7.97×10^9/L, lymphocyte count: 0.53×10^9/L, lymphocyte percentage: 6.70%, C-reactive protein: 217.17 mg/L.

Patient 3: White blood cell count: 9.43×10^9/L, lymphocyte count: 0.69×10^9/L, lymphocyte percentage: 7.40%, C-reactive protein: 202.03 mg/L.

Image Atlas of COVID-19
DOI: https://doi.org/10.1016/B978-0-443-18493-2.00001-2

Patient 4: White blood cell count: 3.29×10^9/L, lymphocyte count: 0.91×10^9/L, lymphocyte percentage: 27.7%, C-reactive protein: 66.07 mg/L.

Patient 5: White blood cell count: 3.44×10^9/L, lymphocyte count: 0.46×10^9/L, lymphocyte percentage: 13.3%, C-reactive protein: 14.9 mg/L.

Patient 6: White blood cell count: 4.59×10^9/L, lymphocyte count: 1.65×10^9/L, lymphocyte percentage: 35.90%, C-reactive protein: 10.14 mg/L.

Radiological findings

Radiological findings of Patients 1–6 are shown in Figs. 1.1–1.6.

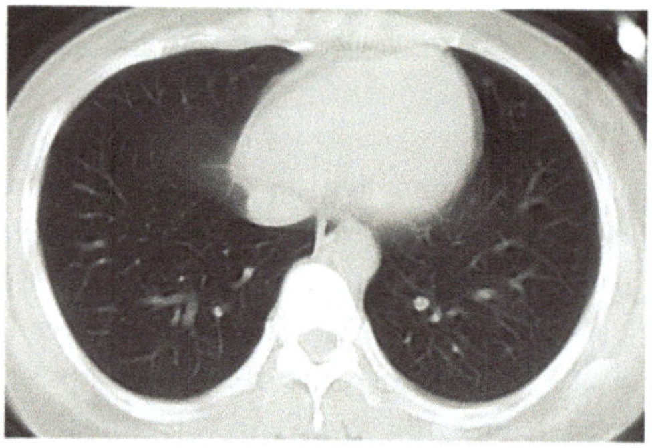

Figure 1.1 Chest computed tomography image of patient 1 shows no remarkable changes.

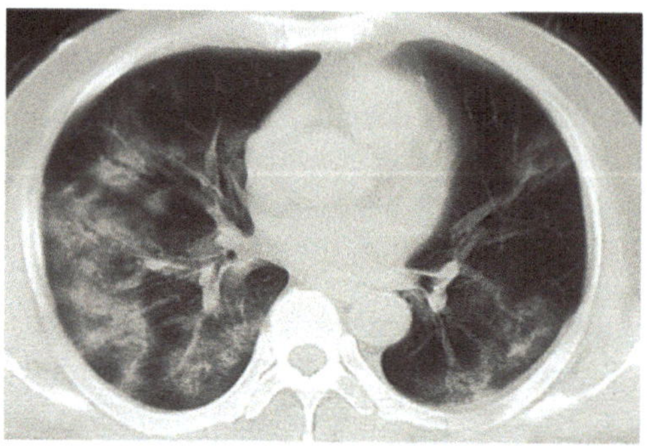

Figure 1.2 Chest computed tomography image of patient 2 shows multiple ground-glass opacities in both lungs.

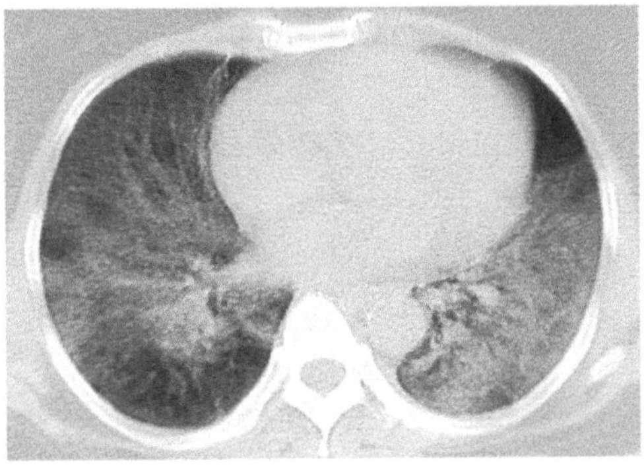

Figure 1.3 Chest computed tomography image of patient 3 shows multiple ground-glass opacities in both lungs.

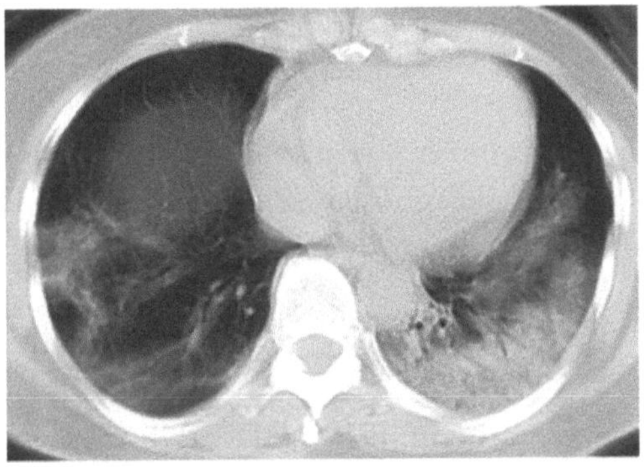

Figure 1.4 Chest computed tomography image of patient 4 shows multiple ground-glass opacities, with partial consolidations and fiber bands seen in the lesions.

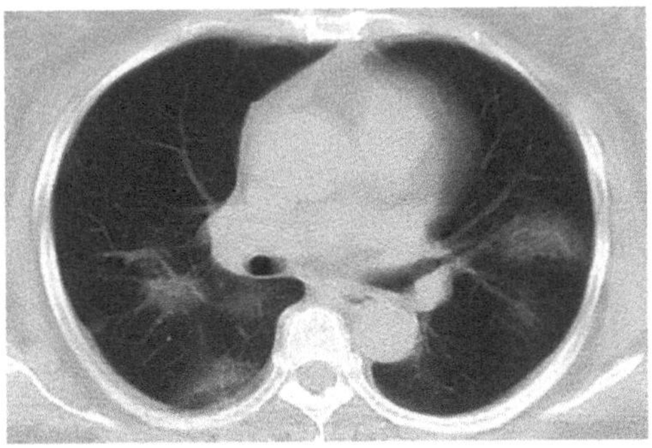

Figure 1.5 Chest computed tomography image of patient 5 shows multiple ground-glass opacities in both lungs.

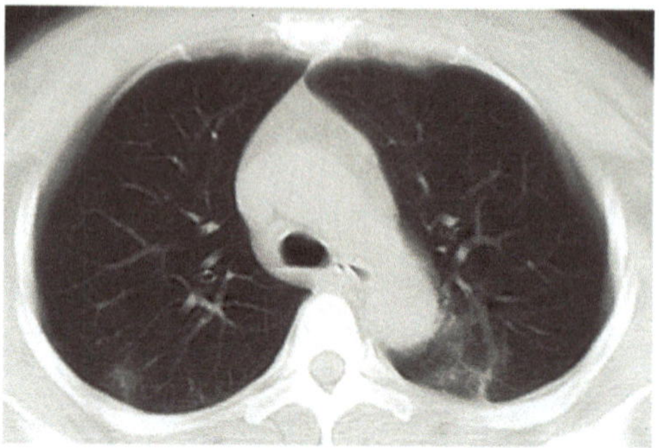

Figure 1.6 Chest computed tomography image of patient 6 shows multiple ground-glass opacities in both lungs.

Section 2: Computed tomography manifestations of COVID-19 cases progressing from mild to common type

Case 7

Medical history

A 51-year-old woman went to her hometown from the epidemic area 4 days prior to the admission. Her husband was diagnosed as COVID-19 positive 3 days prior to the admission. The woman reported throat discomfort 2 h prior to the admission.

Laboratory findings

White blood cell count: 9.20×10^9/L, lymphocyte count: 1.26×10^9/L, lymphocyte percentage: 13.8%, C-reactive protein: 0.7 mg/L.

Radiological findings

The radiological findings of chest computed tomography scans during her hospitalization are as follows: Figs. 1.7–1.10.

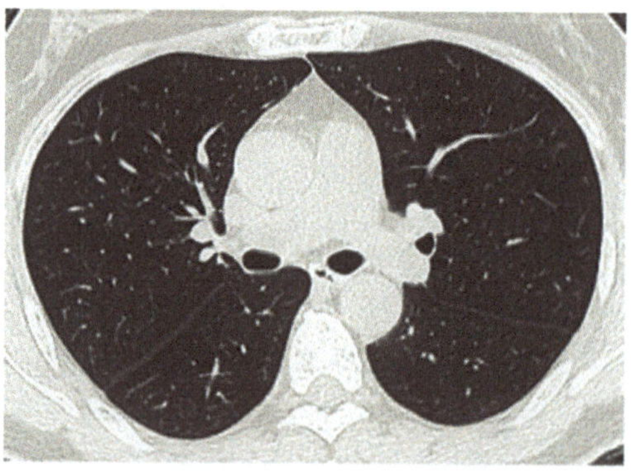

Figure 1.7 Chest computed tomography image on the day of admission. There are no remarkable changes in either lung.

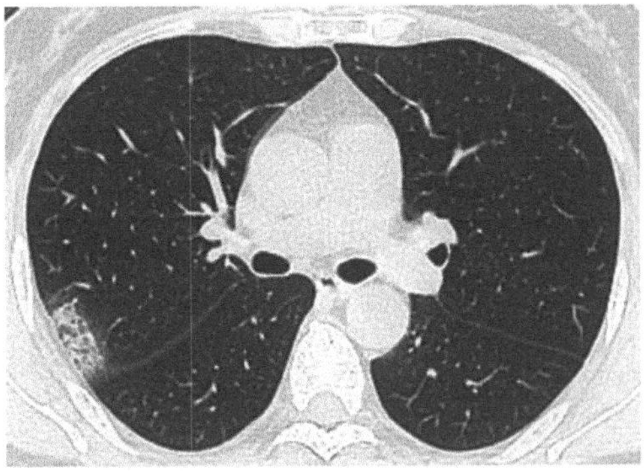

Figure 1.8 Chest computed tomography image 5 days after admission. There is a new emerging ground-glass opacity in the posterior segment of the right lung's upper lobe.

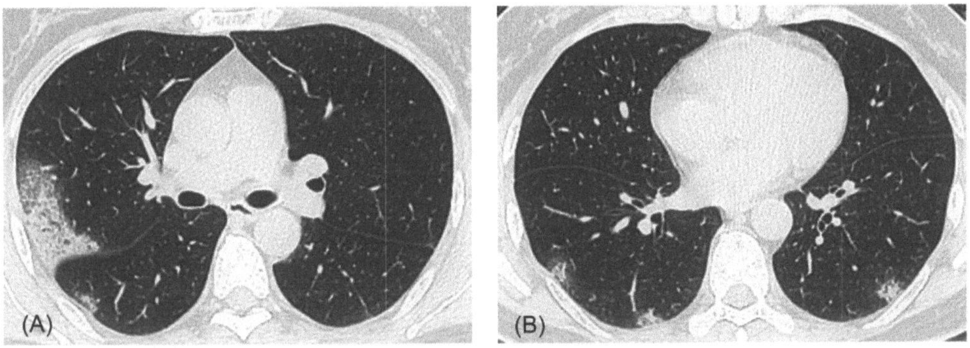

Figure 1.9 Chest computed tomography image 9 days after admission. (A) the ground-glass opacity in the posterior segment of right upper lobe shows enlargement in area, with part of it turning into consolidative opacity, and there is a new emerging ground-glass opacity in the dorsal segment of the right lung's lower lobe; (B) there are multiple new emerging ground-glass opacities in both lungs' lower lobes.

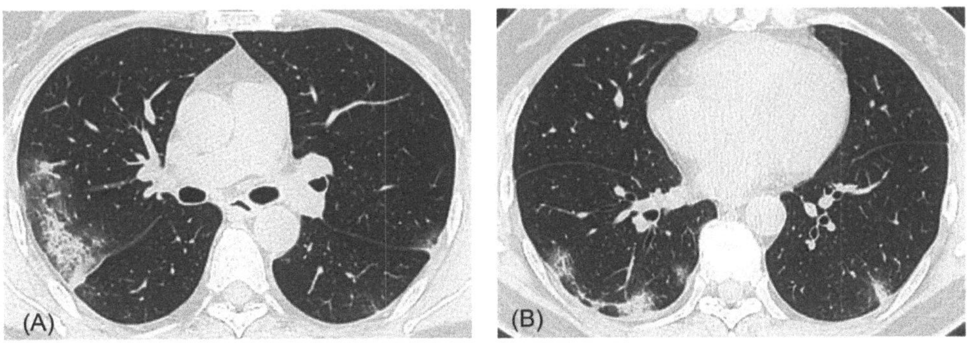

Figure 1.10 Chest computed tomography image 16 days after admission. (A) there are partial absorption for the lesions in the posterior segment of the right lung's upper lobe and complete disappearance of the lesion in the upper segment of the right lung's lower lobe; (B) the lesions' area in the lower lobes of both lungs increases, part of them turning into consolidation, and there are crazy-paving sign and bands seen in the lesions.

Case 8

Medical history

The patient is an 83-year-old woman who came in contact with one of her family members who was diagnosed with COVID-19. She reported fever 1 day prior to the admission with a maximum temperature of 38.2°C, accompanied by cough and dizziness.

Laboratory findings

White blood cell count: $3.70 \times 10^9/L$, lymphocyte count: $1.05 \times 10^9/L$, lymphocyte percentage: 28.2%, C-reactive protein: 5.8 mg/L.

Radiological findings

The radiological findings of chest CT scans during her hospitalization are as follows: Figs. 1.11–1.14.

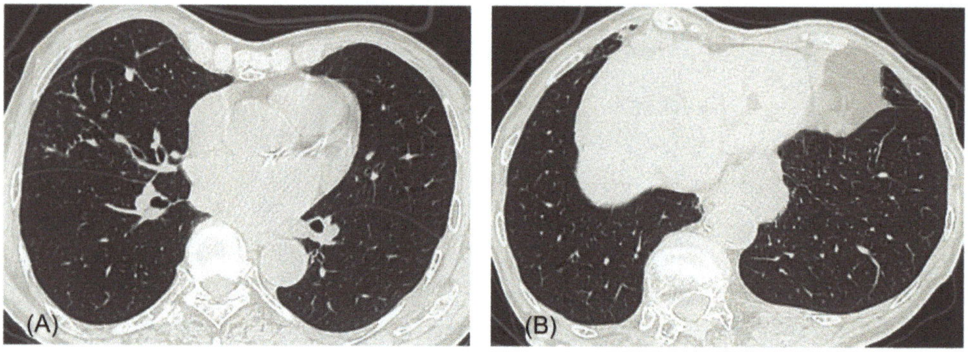

Figure 1.11 Chest computed tomography images on the day of admission. (A) No abnormal findings are noted in the left upper lobe, left lower lobe, right middle lobe and right lower lobe; there are multiple previously existing nodules in the right middle lobe. (B) No abnormal changes are found in the bilateral lower lobes.

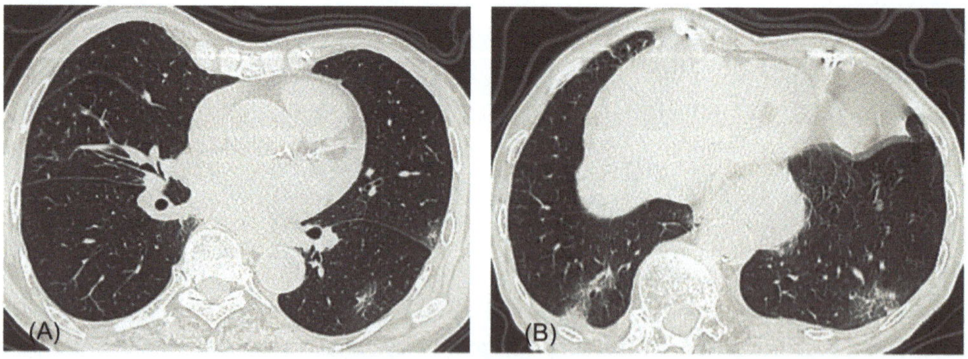

Figure 1.12 Chest computed tomography images 6 days after admission. (A) New emerging ground-glass opacities and consolidations are noted in the left upper lobe and left lower lobe. (B) New ground-glass opacities and consolidations are noted in the bilateral lower lobes, with bronchiolectasis around the lesions.

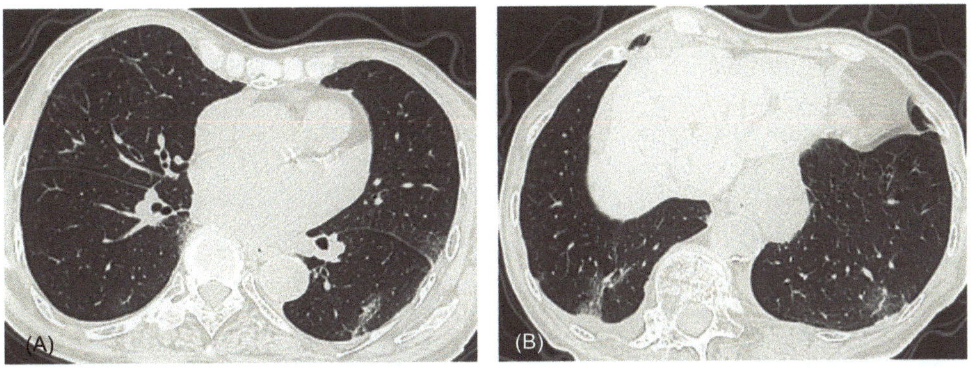

Figure 1.13 Chest computed tomography images 10 days after admission. (A) Partial absorption of lesions is noted in the left upper lobe and left lower lobe. (B) Partial absorption of lesions is noted in the bilateral lower lobes.

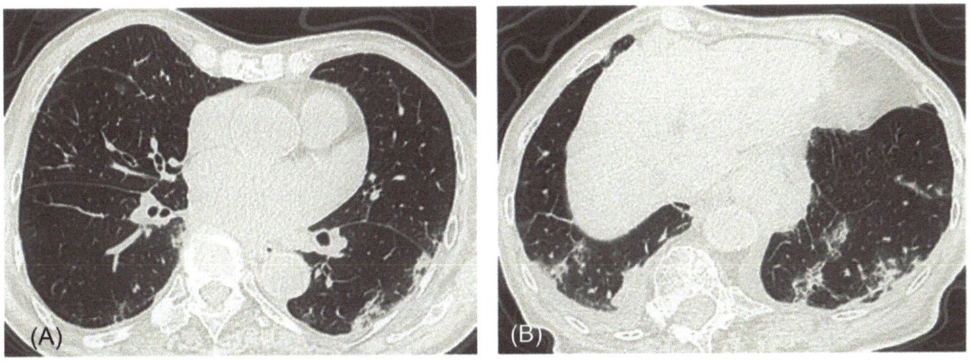

Figure 1.14 Chest computed tomography images 13 days after admission. (A) The lesions in the upper lobe of the left lung almost disappear. New consolidations in the lower lobe of the left lung are noted, and the original consolidations in the lower lobe of the left lung enlarge; there are consolidations in the lower lobe of the right lung. (B) There are new lung consolidations and fibrotic bands in the lower lobes of both lungs.

Case 9

Medical history

The patient is a 28-year-old man. He was documented with a travel history that indicated he visited the epidemic area. He reported an intermittent fever 1 day prior to the admission with a maximum temperature of 37.8°C.

Laboratory findings

White blood cell count: 3.30×10^9/L, lymphocyte count: 1.41×10^9/L, lymphocyte percentage: 43.0%, C-reactive protein: 10.4 mg/L.

Radiological findings

The radiological findings of chest CT scans during his hospitalization are as follows: Figs. 1.15–1.19.

The radiological findings of chest CT scans after discharge are as follows:

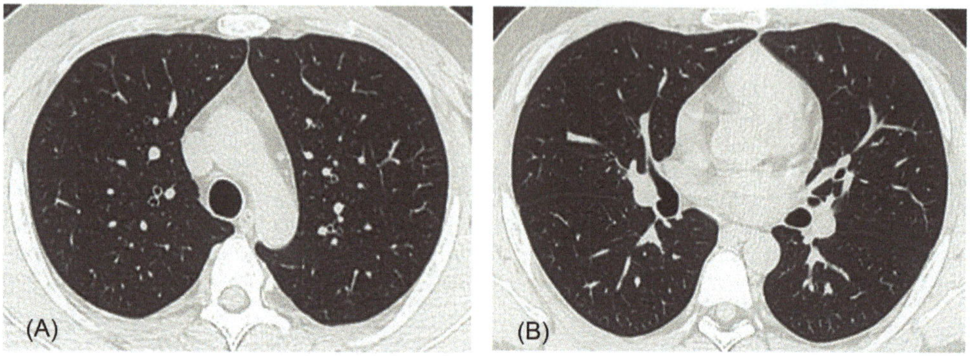

Figure 1.15 Chest computed tomography images on the day of admission. (A) No abnormal changes are noted in the upper lobes of both lungs. (B) No abnormal changes are noted in the upper and lower lobes of both lungs.

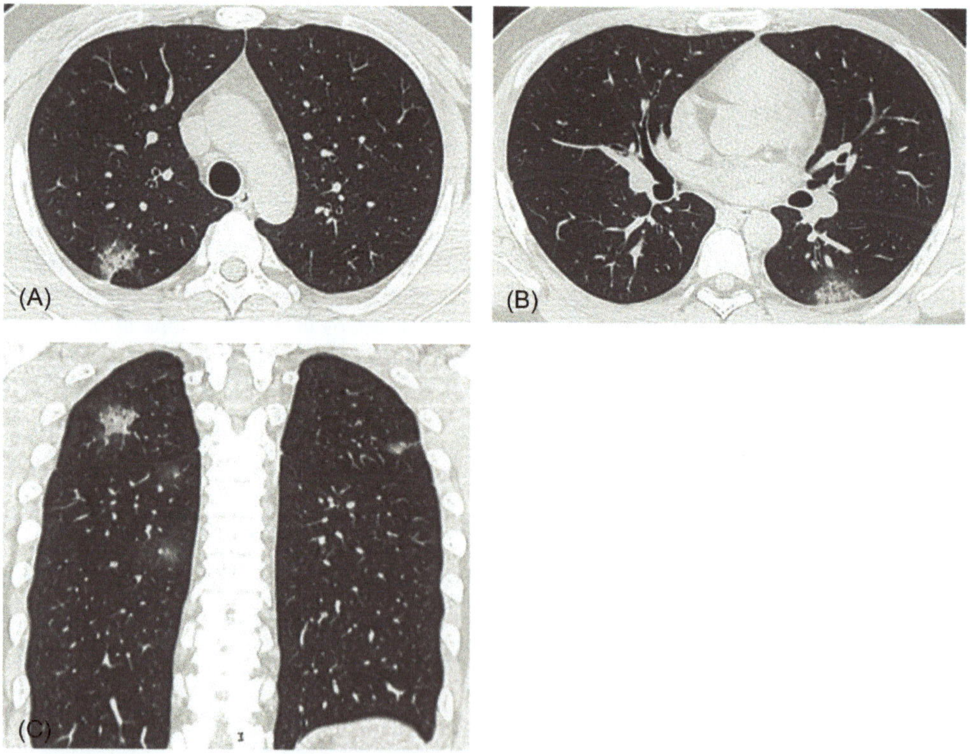

Figure 1.16 Chest computed tomography images 6 days after admission. (A) New ground-glass opacities and consolidations in the upper lobe of the right lung are noted. (B) New ground-glass opacities and consolidations in the dorsal segment of the left lower lobe are noted. (C) The coronal image shows multiple ground-glass opacities and consolidations in the upper lobe of both lungs and the lower lobe of the right lung.

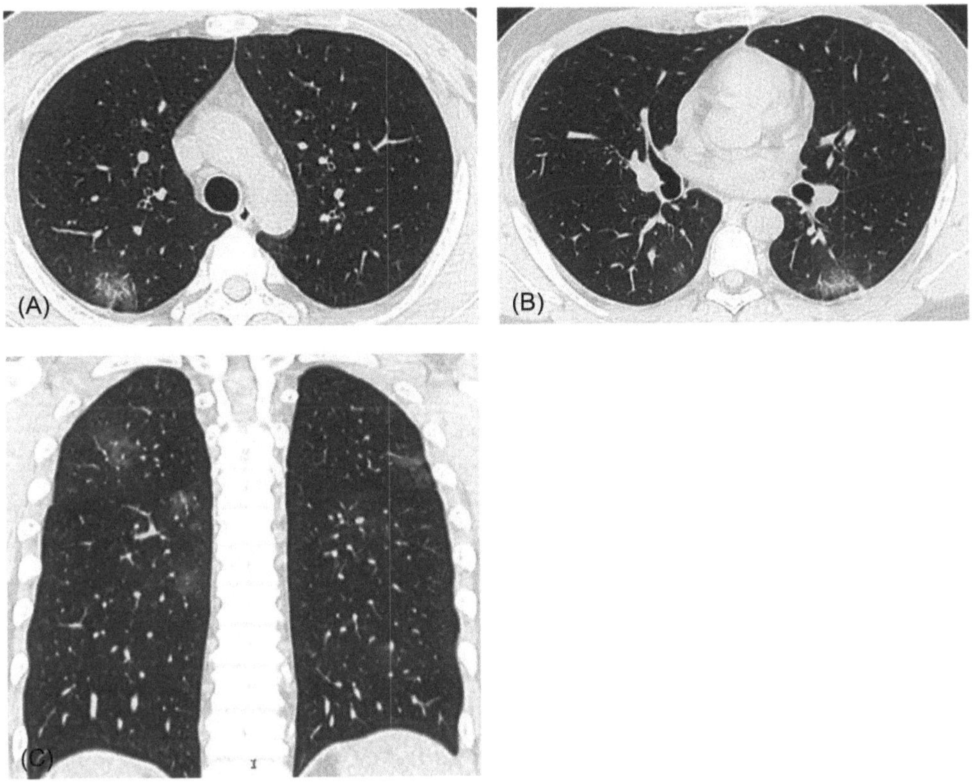

Figure 1.17 Chest computed tomography images 9 days after admission. (A) There is partial absorption of ground-glass opacities and consolidations in the upper lobe of the right lung. (B) There is almost no change with the lesions of the left lower lobe; there is new ground-glass opacity in the right lower lobe. (C) The coronal image shows that the lesions in the upper lobe of the two lungs and the lower lobe of the right lung are partially absorbed.

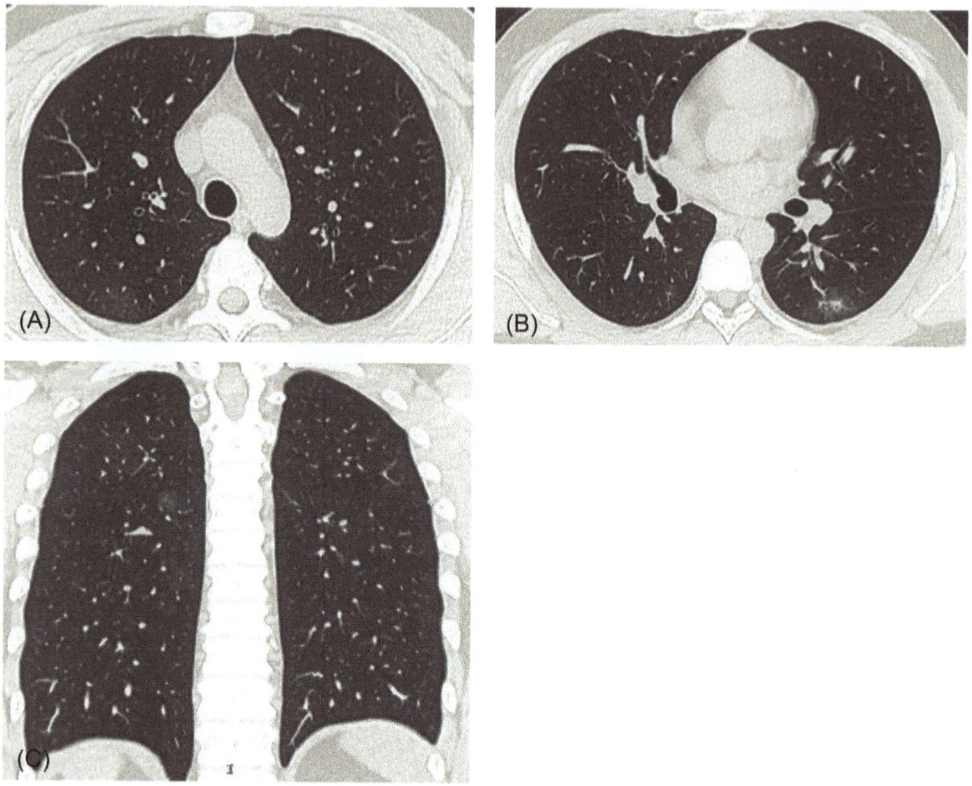

Figure 1.18 Chest computed tomography images 15 days after admission. (A) The lesions in the upper lobe of the right lung continue to be absorbed, with mild ground-glass opacity left. (B) The lesions in the left lower lobe and right lower lobe lesions show absorption. (C) The coronal image shows that the lesions in the upper lobes of the two lungs are basically absorbed, and the lesions in the lower lobe of the right lung are absorbed.

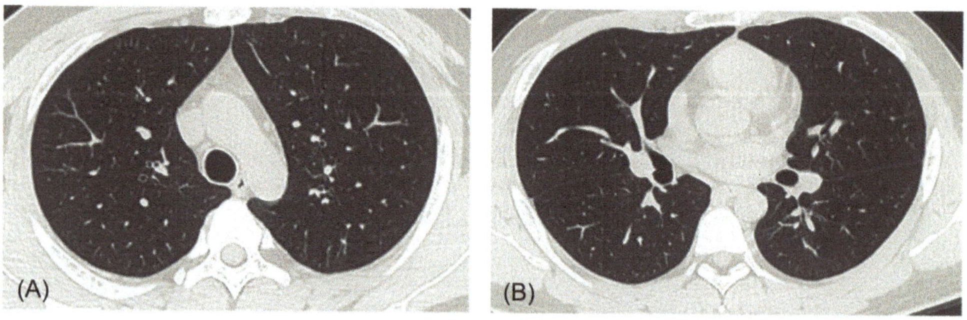

Figure 1.19 Chest computed tomography images 1 month after discharge. (A) The lesions in the upper lobes of both lungs disappear. (B) The lesions in bilateral lower lobes disappear.

Reference

[1] Bai Y, Yao L, Wei T, Tian F, Jin D-Y, Chen L, et al. Presumed asymptomatic carrier transmission of COVID-19. JAMA 2020;323(14):1406–7.

CHAPTER 2

Computed tomography manifestations and prognosis of common COVID-19 cases

Case 10

Medical history

A 22-year-old man who was staying in the epidemic area left the place just 6 days prior to the admission. He reported a fever 3 days prior to the admission with a maximum temperature of 37.8°C, accompanied by a cough and a runny nose.

Laboratory findings

White blood cell count: 2.90×10^9/L, lymphocyte count: 1.13×10^9/L, lymphocyte percentage: 38.5%, C-reactive protein: 3.4 mg/L.

Radiological findings

The radiological findings of the chest computed tomography (CT) scan during his hospitalization are as follows: Figs. 2.1 and 2.2.

Case 11

Medical history

A 42-year-old woman who was staying in the epidemic area left the place just 2 days prior to the admission. She reported a fever 8 days prior to the admission with a maximum temperature of 37.3°C, accompanied by muscle ache and headache. The symptoms did not go away after 5 days of treatment. The temperature was still 37.2°C with dry cough, dizziness, a runny nose, fatigue, and occasional asthma.

Laboratory findings

White blood cell count: 5.23×10^9/L, lymphocyte count: 1.29×10^9/L, lymphocyte percentage: 24.7%, C-reactive protein: 1.24 mg/L.

Image Atlas of COVID-19
DOI: https://doi.org/10.1016/B978-0-443-18493-2.00002-4

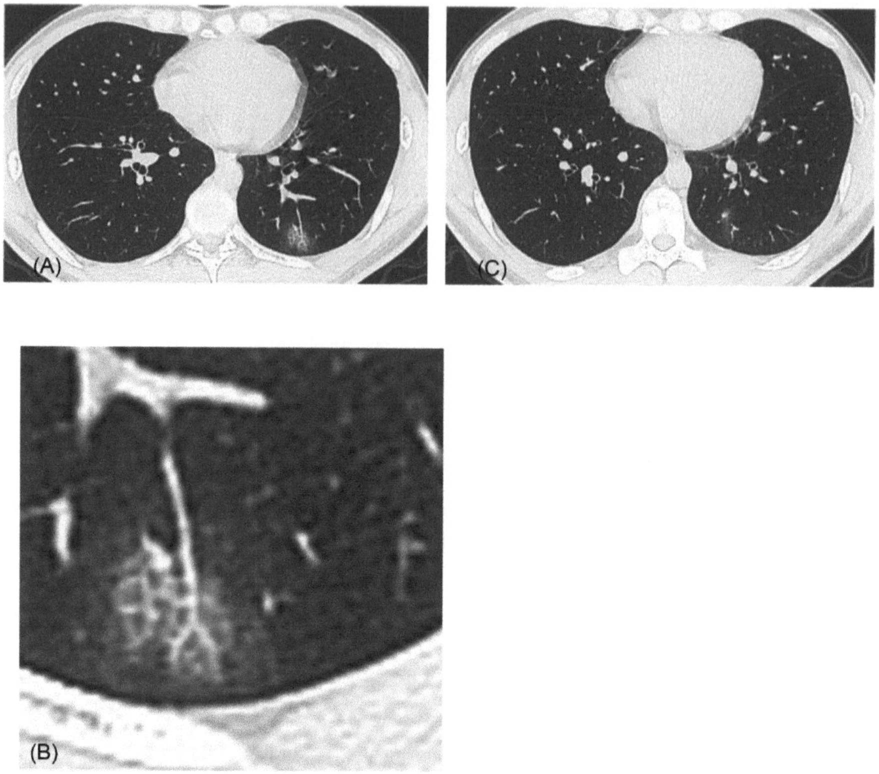

Figure 2.1 Chest computed tomography images on the day of admission. (A) There is a ground-glass opacity in the left lung's lower lobe; (B) the intra- and interlobular septa are thickened (crazy-paving sign), and the vessels in the lesion are dilated; (C) there are ground-glass and consolidative opacities in the left lung's lower lobe.

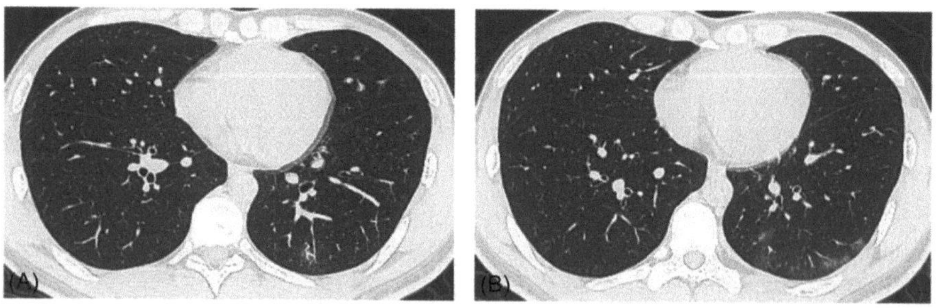

Figure 2.2 Chest computed tomography images 6 days after admission. (A) The ground-glass opacity in the left lung's lower lobe shows significant absorption, with remaining mild bands; (B) there are several new emerging ground-glass opacities in the left lung's lower lobe.

Radiological findings

The radiological findings of the chest CT scan during her hospitalization are as follows: Figs. 2.3–2.5.

The radiological findings of chest CT scan after discharge are as follows: Fig. 2.6.

Case 12

Medical history

A 53-year-old man lives in the epidemic area. He reported a fever 1 day prior to the admission with a maximum temperature of 37.7°C, accompanied by dizziness.

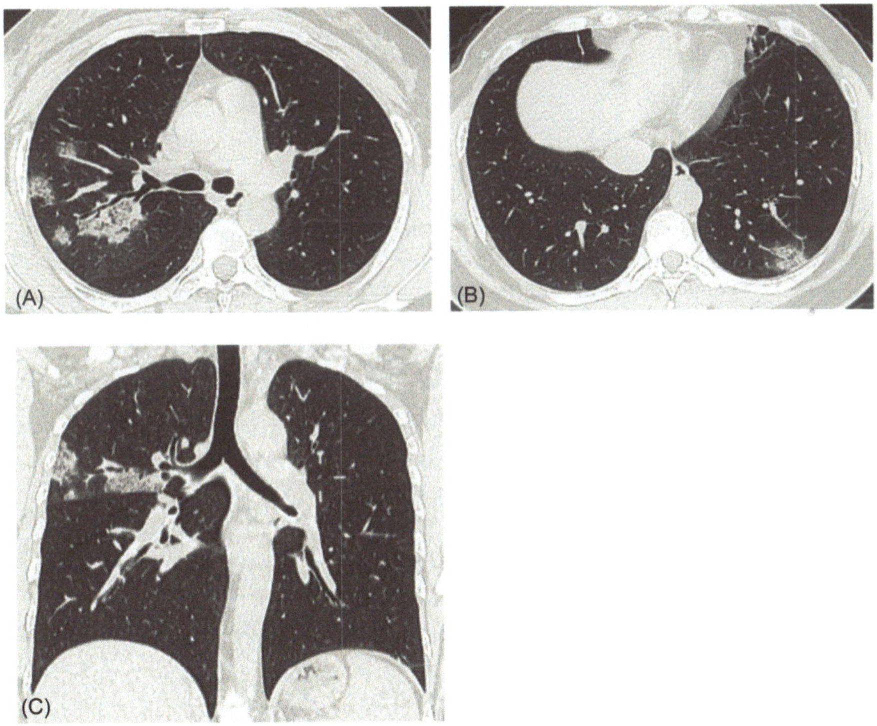

Figure 2.3 Chest computed tomography images on the day of admission. (A) There are multiple ground-glass and consolidative opacities in the right lung's upper lobe; (B) ground-glass, consolidative opacities and crazy-paving sign are visible in the peripheral area of the posterior basal segment of the left lung's lower lobe; (C) coronal image shows the lesions in the right lung's upper lobe distributed along the bronchus.

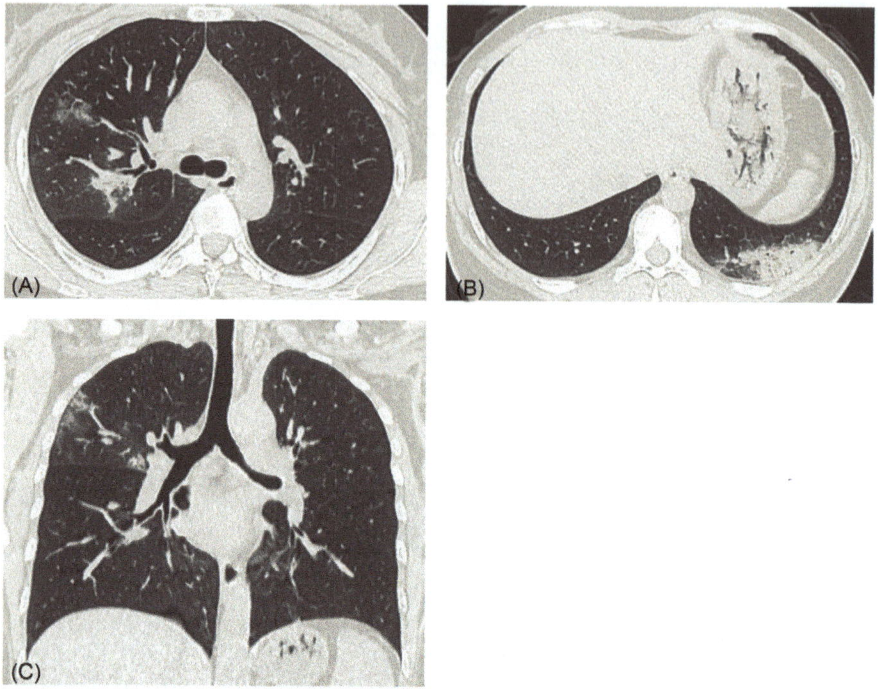

Figure 2.4 Chest computed tomography images 10 days after admission. (A) There is obvious absorption of ground-glass and consolidative opacities in the right lung's upper lobe, with there remaining only bands and mild ground-glass opacities; (B) the lesions in the posterior basal segment of the left lung's lower lobe enlarge significantly; (C) coronal image shows obvious reduction for lesions in the right lung's upper lobe.

Laboratory findings

White blood cell count: 3.40×10^9/L, lymphocyte count: 1.03×10^9/L, lymphocyte percentage: 30.3%, C-reactive protein: 47.7 mg/L.

Radiological findings

The radiological findings of the chest CT scan during his hospitalization are as follows: Figs. 2.7–2.10

Case 13

Medical history

A 34-year-old woman who lives in the epidemic area left the place just 4 days prior to the admission. She reported a fever 3 days prior to the examination of her

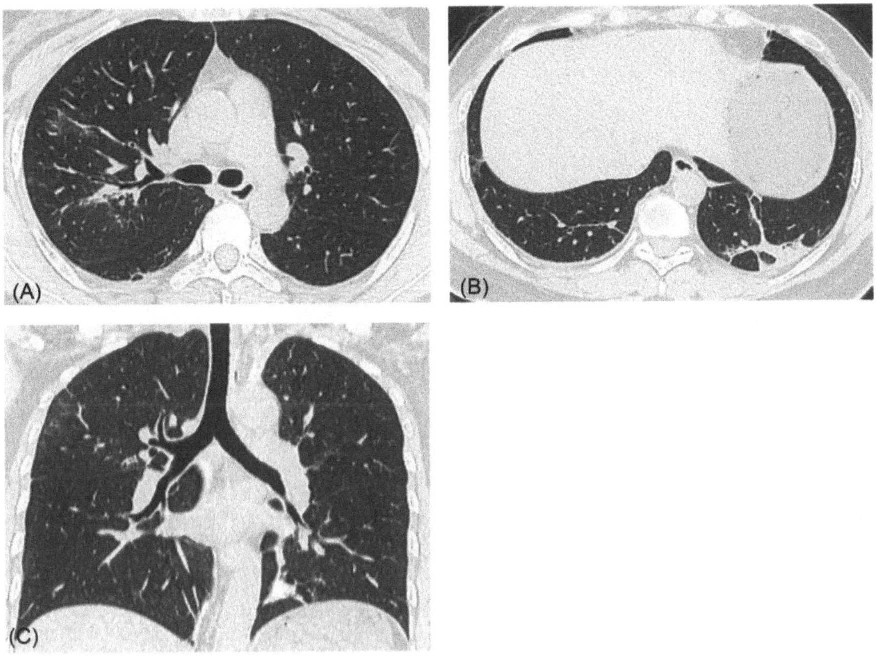

Figure 2.5 Chest computed tomography images 14 days after admission. (A) The absorption continues in the right lung's upper lobe, with there remaining only mild ground-glass opacities; (B) the lesions in the posterior basal segment of the left lung's lower lobe reduce significantly, with there remaining only parenchymal bands; (C) coronal image shows obvious reduction for lesions in the posterior segment of the right lung's upper lobe.

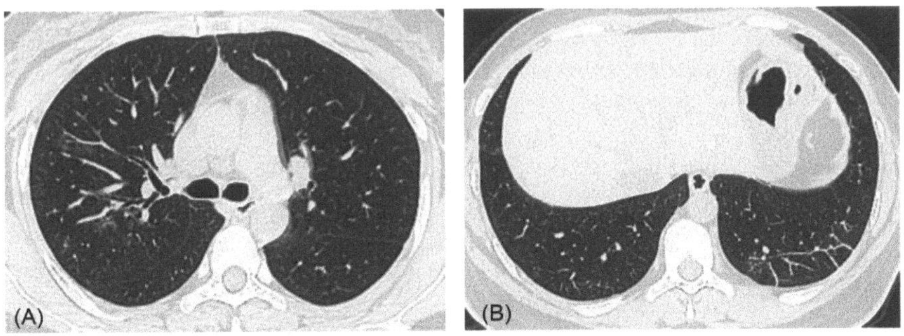

Figure 2.6 Chest computed tomography images 17 days after discharge. (A) The absorption continues in the right lung's upper lobe, with there remaining only mild ground-glass opacities. (B) The lesions in the posterior basal segment of the left lung's lower lobe reduce significantly, with there remaining only parenchymal bands.

condition with a maximum temperature of 39.0°C, accompanied by chills, headaches, weakness, and muscle soreness. She reported a dry cough and palpitation 1 day prior to the admission.

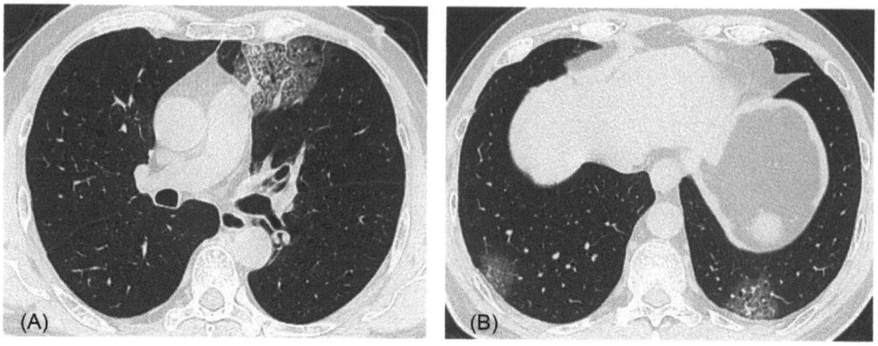

Figure 2.7 Chest computed tomography images on the day of admission. (A) There are ground-glass opacities in the upper lobe of the left lung, with air bronchograms and crazy-paving sign seen inside. (B) Ground-glass opacities are seen in the lower lobes of both lungs, with air bronchograms seen inside.

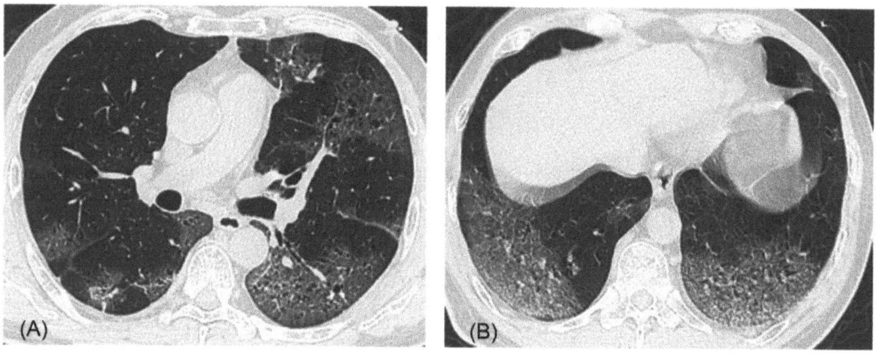

Figure 2.8 Chest computed tomography images 6 days after admission. (A) Multiple new ground-glass opacities are seen in the upper and lower lobes of both lungs, with dilated vessels and bronchiolectasis seen inside. (B) The ground-glass opacities in the lower lobes of both lungs enlarge.

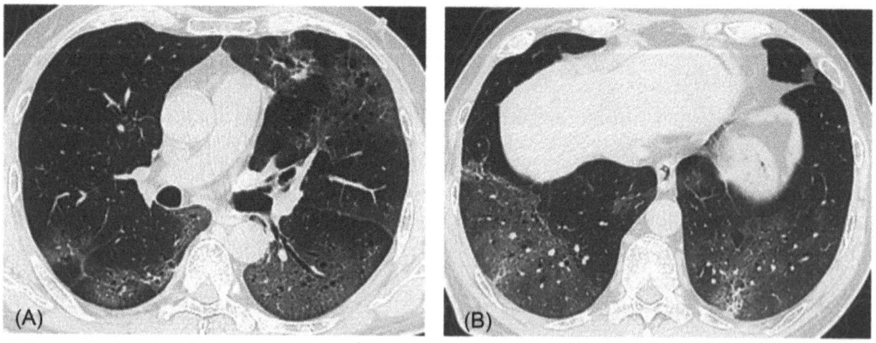

Figure 2.9 Chest computed tomography images 10 days after admission. (A) The lesions in the upper and lower lobes of both lungs show absorption. (B) The attenuation of the lesions in the lower lobes of the two lungs become lighter, with consolidations and bands remaining in them.

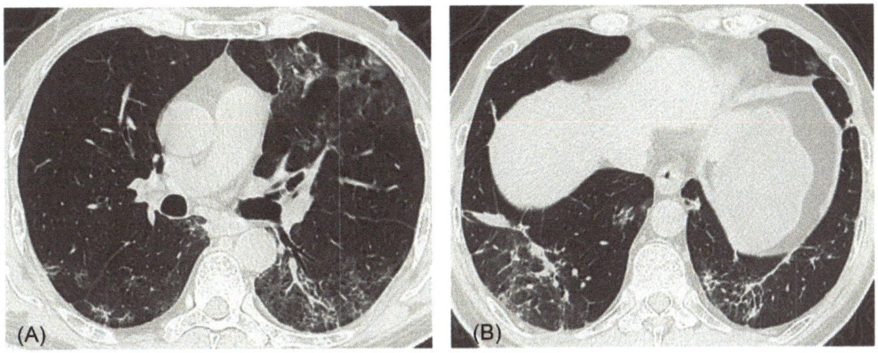

Figure 2.10 Chest computed tomography images 19 days after admission. (A) The lesions in the upper and lower lobes of both lungs show absorption, with residual consolidations and bands. (B) The lesions in the lower lobes of both lungs show absorption.

Laboratory findings

White blood cell count: 2.68×10^9/L, lymphocyte count: 0.87×10^9/L, lymphocyte percentage: 32.5%, C-reactive protein: 7.5 mg/L.

Radiological findings

The radiological findings of the chest CT scan during her hospitalization are as follows: Figs. 2.11 and 2.12.

Case 14

Medical history

A 34-year-old man who lives in the epidemic area left the place just 4 days prior to the admission. He reported fever during admission to the hospital with a maximum temperature of 39.0°C, accompanied by weakness and muscle soreness.

Laboratory findings

White blood cell count: 4.50×10^9/L, lymphocyte count: 1.40×10^9/L, lymphocyte percentage: 31.3%, C-reactive protein: 21.1 mg/L.

Radiological findings

The radiological findings of the chest CT scan during his hospitalization are as follows: Figs. 2.13–2.16.

The radiological findings of chest CT scan after discharge are as follows: Fig. 2.17.

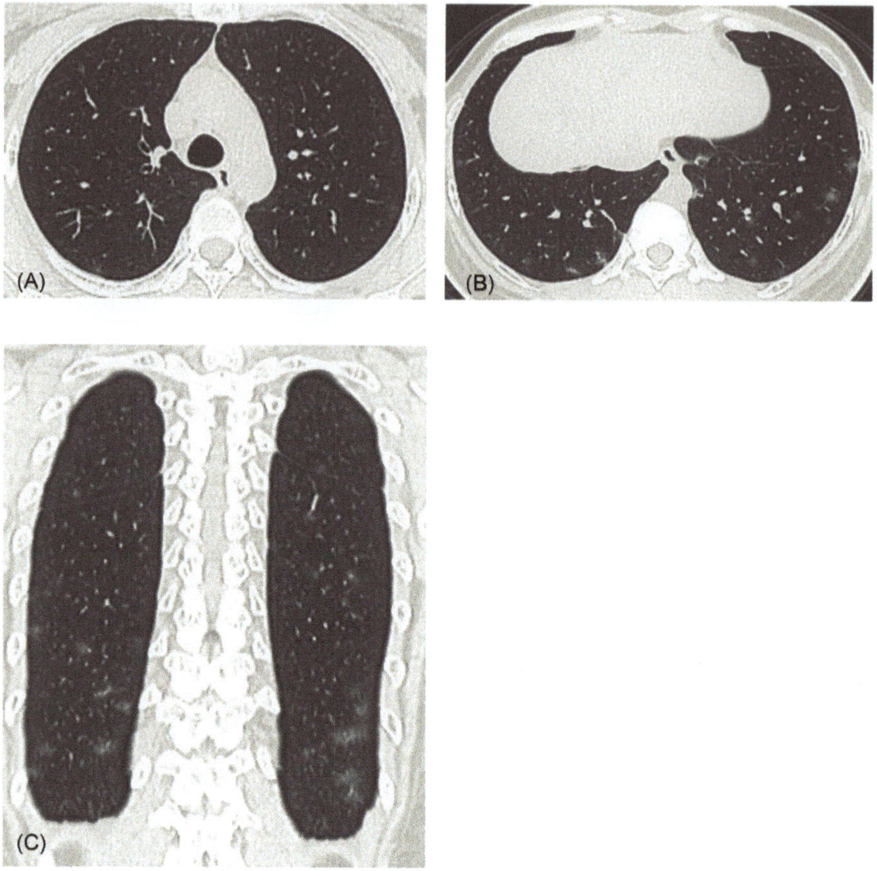

Figure 2.11 Chest computed tomography images on the day of admission. (A) Multiple ground-glass opacities are seen in the subpleural area of the bilateral upper lobes. (B) Multiple ground-glass opacities are seen in the subpleural area of the bilateral lower lobes. (C) The coronal image shows multiple ground-glass opacities in the upper and lower lobes of both lungs, predominantly in the lower lobes of the two lungs.

Case 15

Medical history

A 35-year-old man documented a history of travel to the epidemic area. He reported a fever 9 days prior to the admission with a maximum temperature of 37.3°C, accompanied by fatigue. He started to have chest tightness and cough 5 days prior to the admission. The throat swab nucleic acid test was positive 2 days prior to the admission.

Laboratory findings

White blood cell count: 3.20×10^9/L, lymphocyte count: 0.66×10^9/L, lymphocyte percentage: 20.3%, C-reactive protein: 11.2 mg/L.

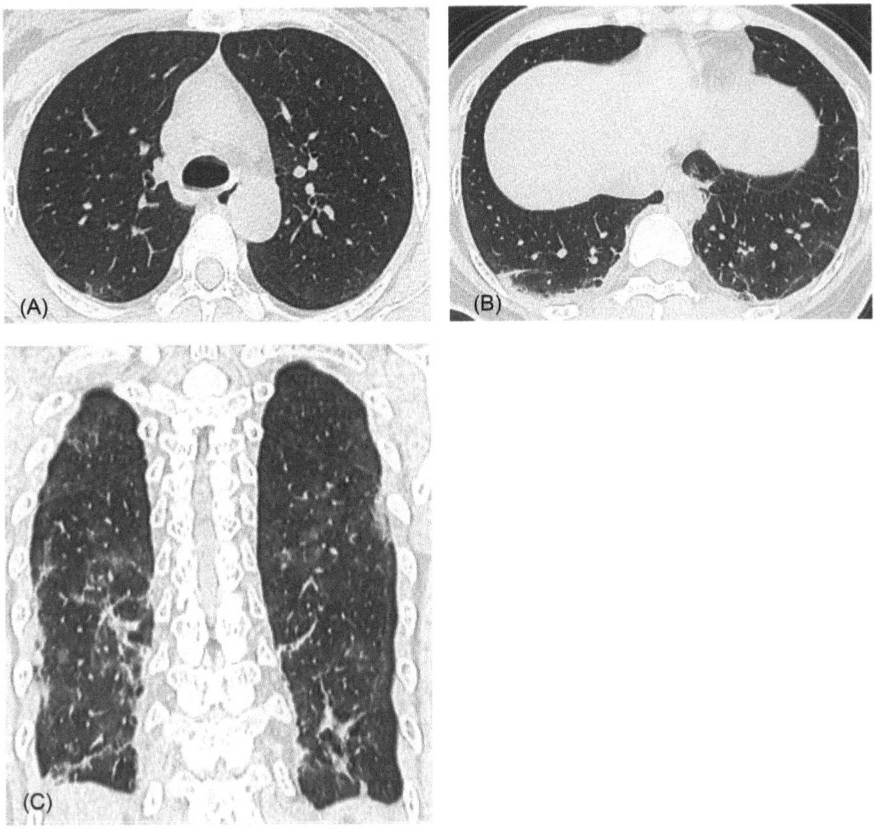

Figure 2.12 Chest computed tomography images 6 days after admission. (A) The lesions enlarge in the bilateral upper lobes. (B) The lesions enlarge in the bilateral lower lobes, with consolidations and bands seen inside. (C) The coronal image shows multiple ground-glass opacities in both lungs, with consolidations and bands mainly in the lower lobes.

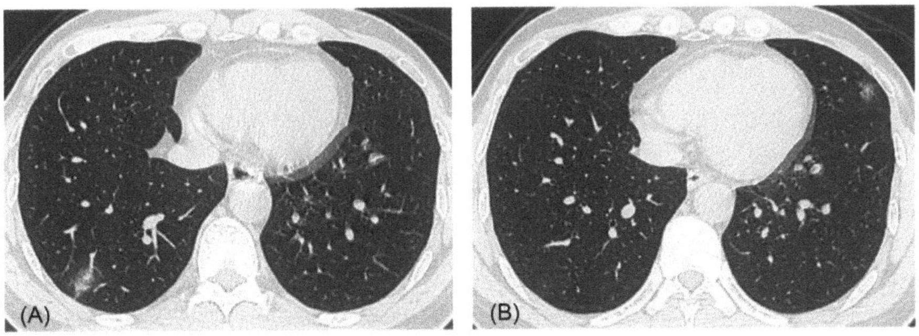

Figure 2.13 Chest computed tomography images on the day of admission. (A) There are ground-glass and consolidative opacities in the right lung's lower lobe. (B) There is ground-glass opacity in the left lung's upper lobe.

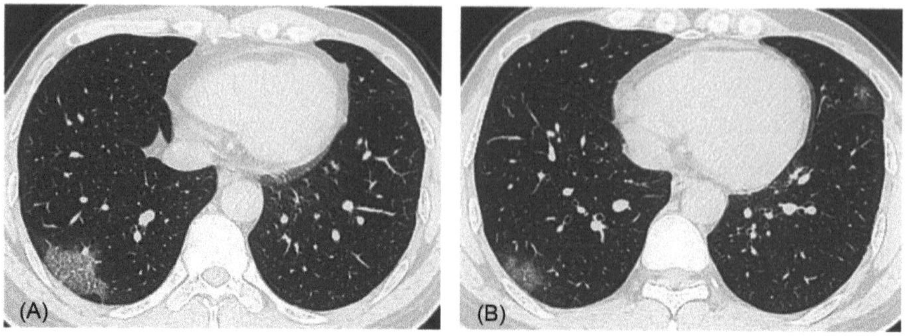

Figure 2.14 Chest computed tomography images 4 days after admission. (A) The lesions enlarge in the right lung's lower lobe, with crazy-paving sign seen in it. (B) The lesion in the left lung's upper lobe shows slight absorption, with a reduction in the attenuation.

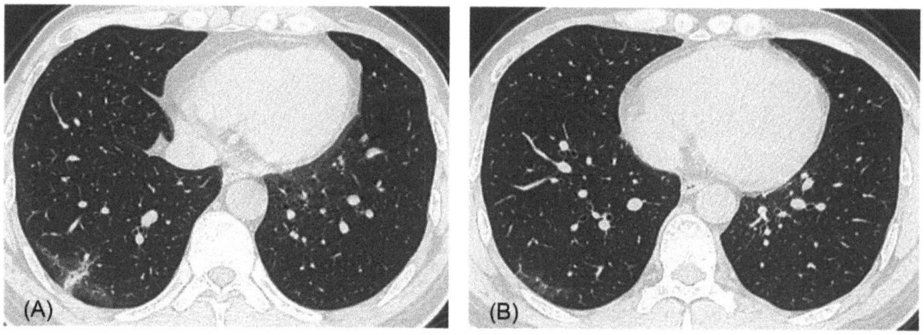

Figure 2.15 Chest computed tomography images 9 days after admission. (A) There is obvious absorption for the lesions in the right lung's lower lobe, with some ground-glass opacities turning into consolidation and bands; (B) The lesion in the left lung's upper lobe has disappeared.

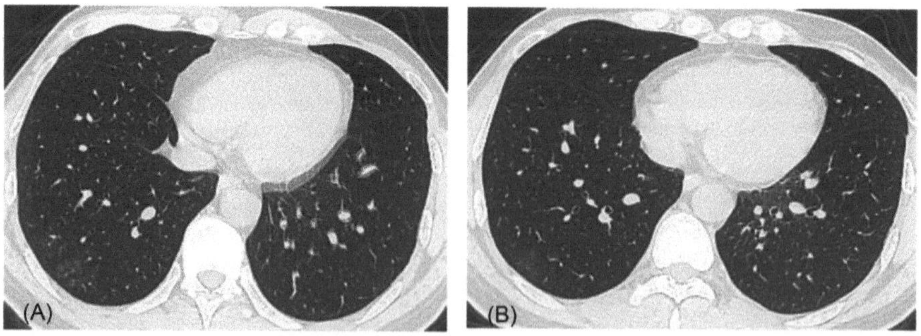

Figure 2.16 Chest computed tomography images one month after discharge. (A) The lesions in the right lung's lower lobe have almost disappeared completely, with there remaining only ground-glass opacities. (B) The lesion in the left lung's upper lobe has disappeared.

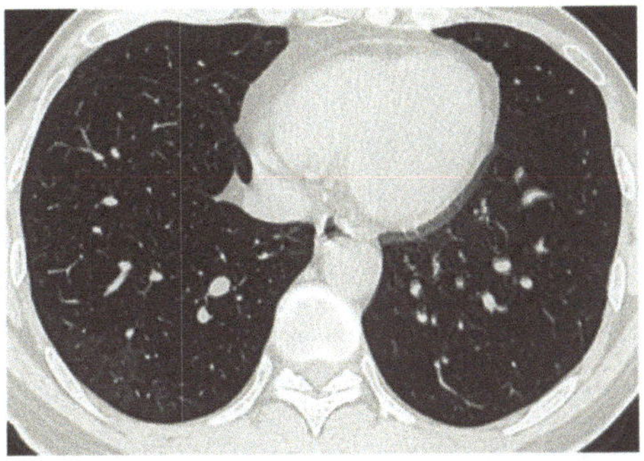

Figure 2.17 Chest computed tomography image one and a half months after discharge. Axial chest computed tomography image shows residual ground-glass opacity in the right lung's lower lobe.

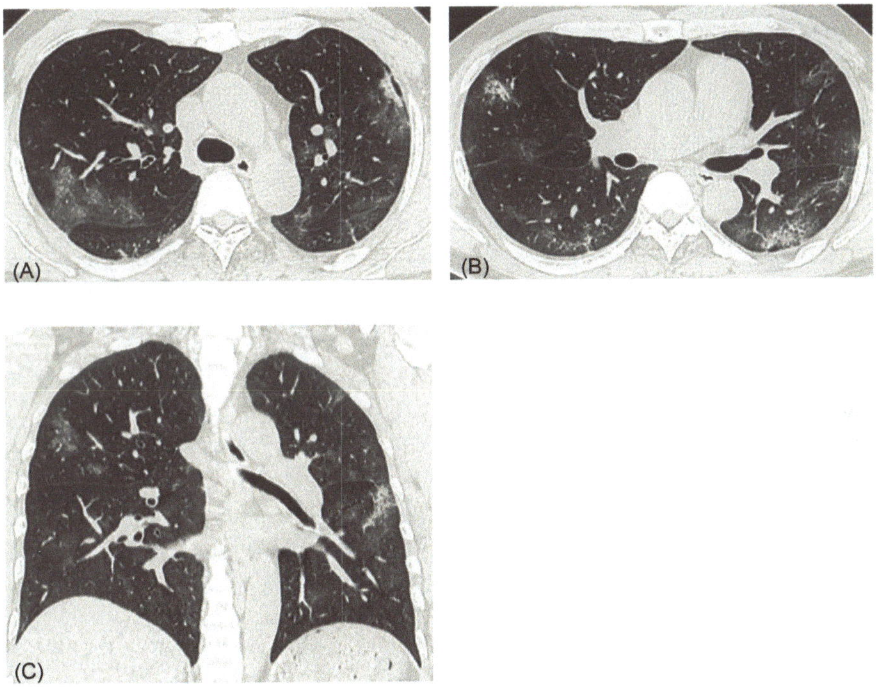

Figure 2.18 Chest computed tomography images 3 days after admission. (A) Multiple ground-glass opacities and consolidations are seen in the upper and lower lobes of both lungs. (B) Multiple ground-glass opacities and consolidations are seen in the bilateral upper and lower lobes, and right middle lobe, with air bronchogram seen in the lesions. (C) The coronal image shows multiple ground-glass opacities in the upper and lower lobes of both lungs, and consolidations in the lower lobe of the left lung.

Radiological findings

The radiological findings of the chest CT scan during his hospitalization are as follows: Figs. 2.18–2.20

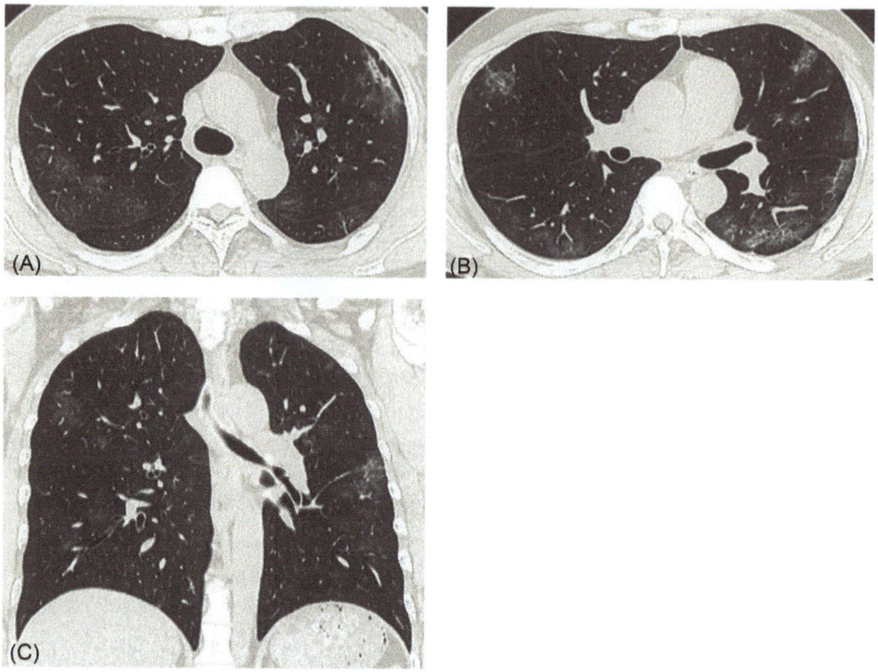

Figure 2.19 Chest computed tomography images 7 days after admission. (A) The lesions of the upper and lower lobes of both lungs show absorption. (B) The lesions in the upper lobe, lower lobe and right middle lobe of both lungs show absorption. (C) The coronal image shows the absorption of lesions in the upper and lower lobes of both lungs.

Case 16

Medical history

The patient is a 57-year-old woman. She lives with her daughter and son, who have both had contact with people who came from the epidemic area. She reported a fever 4 days prior to the admission with a maximum temperature of 38.0°C, accompanied by a cough with white sticky sputum, general weakness, and muscle ache.

Laboratory findings

White blood cell count: 2.56×10^9/L, lymphocyte count: 0.85×10^9/L, lymphocyte percentage: 33.2%, C-reactive protein: 4.62 mg/L.

Radiological findings

The radiological findings of the chest CT scan during her hospitalization are as follows: Figs. 2.21–2.24.

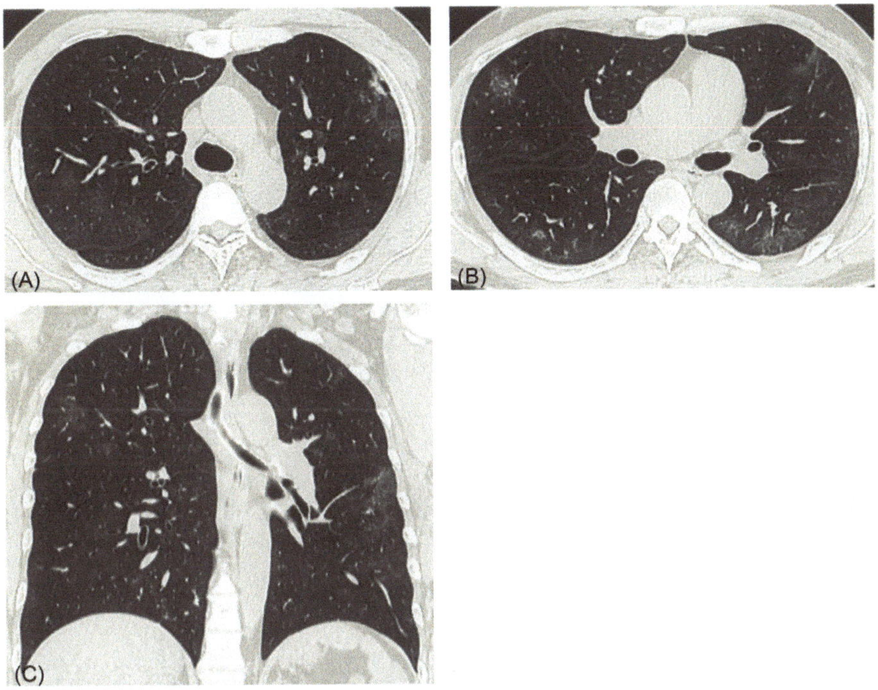

Figure 2.20 Chest computed tomography images 10 days after admission. (A) The lesions in the upper lobes of the two lungs show absorption, and the lesions in the dorsal segment of the lower lobes of the two lungs disappear. (B) The lesions in the upper lobe, lower lobe and right middle lobe of both lungs show absorption. (C) The coronal image shows the absorption of lesions in the upper and lower lobes of both lungs.

Case 17

Medical history

A 54-year-old man reported a fever 1 day prior to the admission after having come in contact with people who came from the epidemic area (diagnosed as COVID-19) with a maximum temperature of 37.7°C, accompanied by fatigue and headache.

Laboratory findings

White blood cell count: 6.40×10^9/L, lymphocyte count: 1.14×10^9/L, lymphocyte percentage: 17.8%, C-reactive protein: 46.3 mg/L.

Radiological findings

The radiological findings of the chest CT scan during his hospitalization are as follows: Figs. 2.25–2.28.

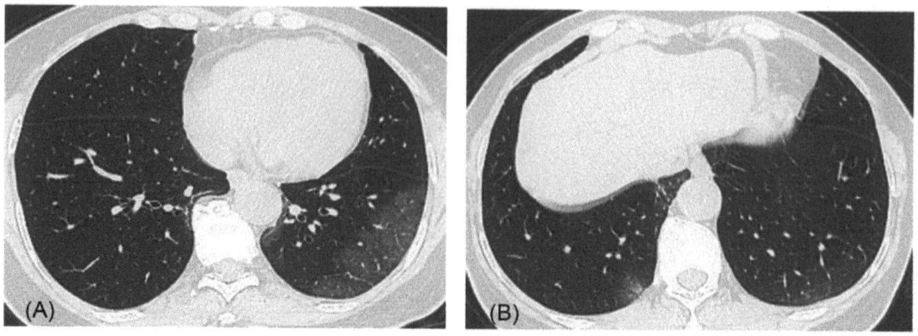

Figure 2.21 Chest computed tomography images on the day of admission. (A) There is a ground-glass opacity in the left lower lobe. (B) There are ground-glass opacities in the subpleural area of the bilateral lower lobes.

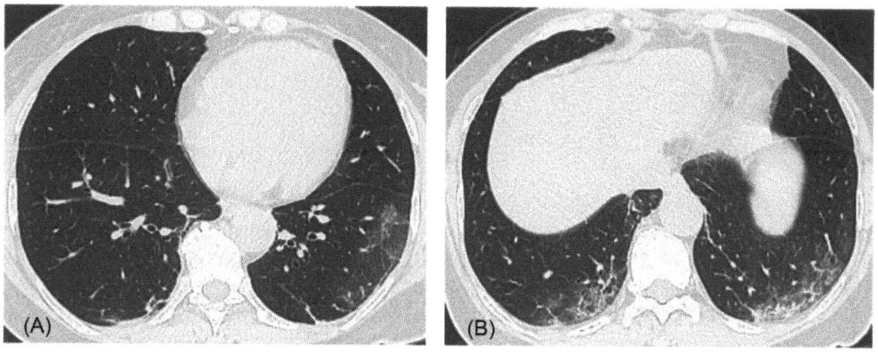

Figure 2.22 Chest computed tomography images 6 days after admission. (A) The lesions in the left lung's lower lobe show mild absorption for the ground-glass opacities, with some of them turning into consolidation and bands. (B) The lesions in the right lung's lower lobe show mild enlargement, with some ground-glass opacities turning into consolidation and bands.

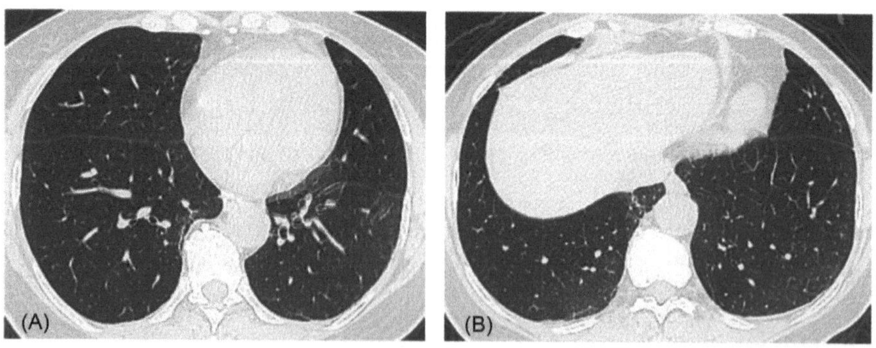

Figure 2.23 Chest computed tomography images 12 days after admission. (A) The lesions in the lower lobes of both lungs show absorption. (B) The lesions in the left lower lobe disappear, with mild bands remaining in the right lower lobe.

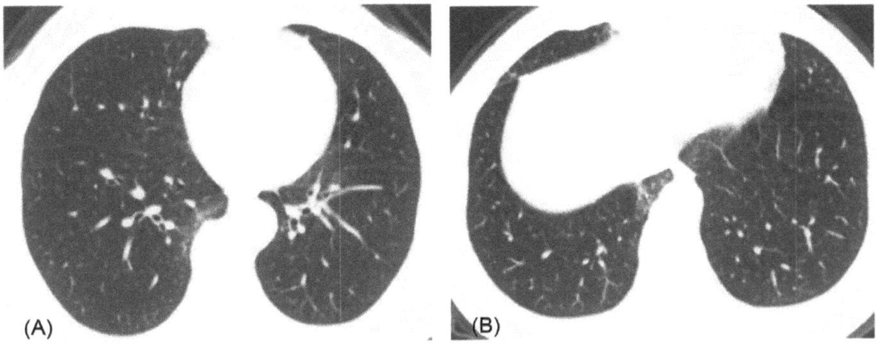

Figure 2.24 Chest computed tomography images one and a half months after discharge. (A) No abnormal findings in both lungs are noted. (B) No abnormal findings in both lungs are noted.

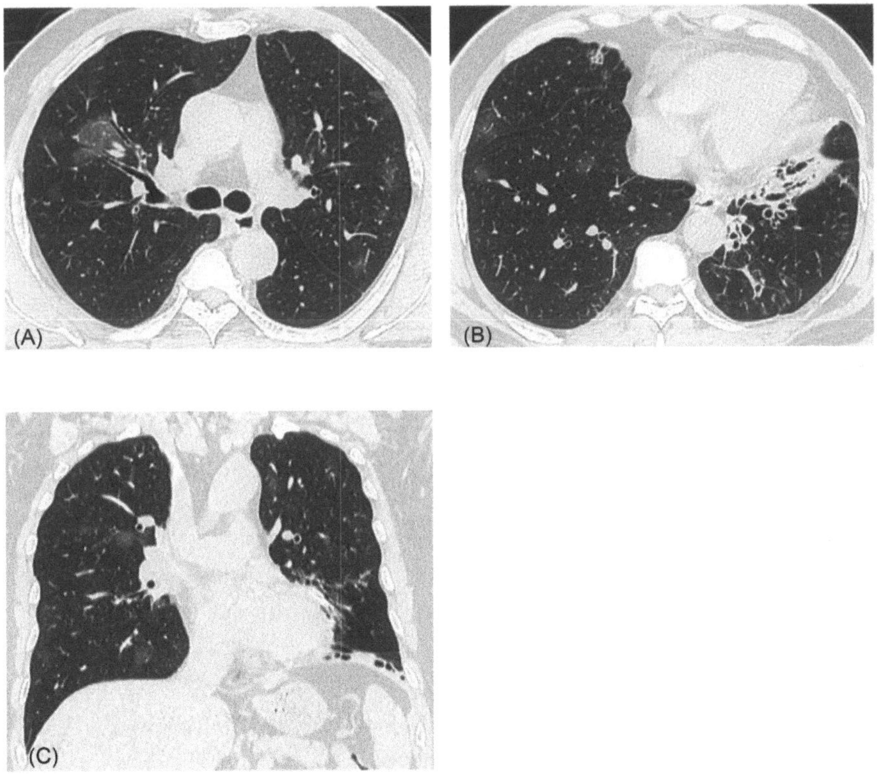

Figure 2.25 Chest computed tomography images on the day of admission. (A) Multiple ground-glass opacities in the upper lobes of both lungs are noted. (B) Multiple ground-glass opacities in the right middle lobe and lower lobe, multiple bronchiectasis and consolidations in the left lower lobe are noted. (C) The coronal image shows multiple ground-glass opacities in bilateral lungs, and bronchiolectasis and consolidations in the lower lobe of the left lung.

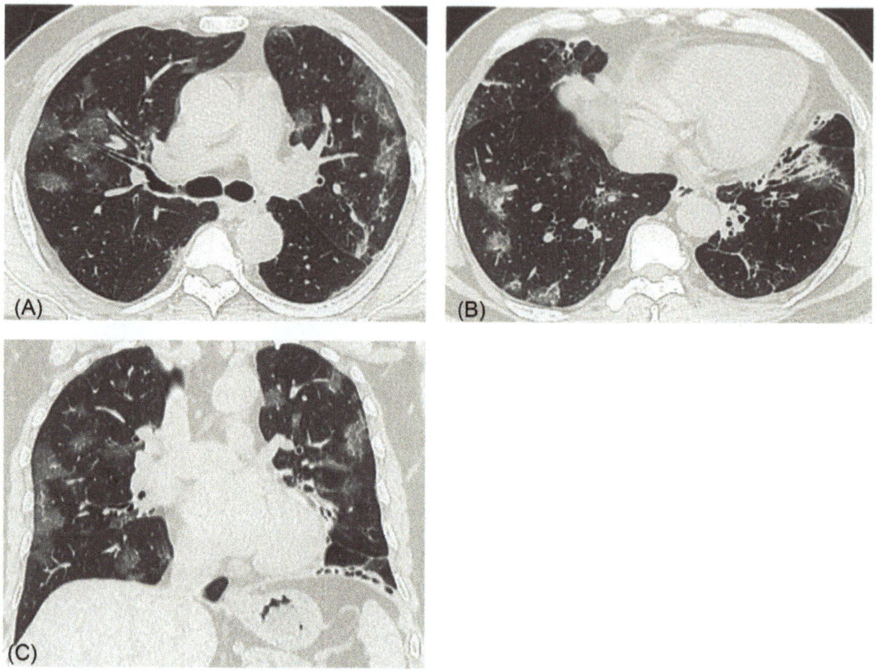

Figure 2.26 Chest computed tomography images 7 days after admission. (A) The ground-glass opacities in the bilateral lungs show significant enlargement both in size and attenuation. (B) The ground-glass opacities in the bilateral lungs show significant enlargement both in size and attenuation. The consolidation and bronchiectasis in the right lung's lower lobe are similar to prior images. (C) The coronal image shows enlargement of ground-glass opacities in bilateral lungs.

Case 18

Medical history

A 28-year-old man documented a history of contact with a person who was diagnosed with COVID-19 just 8 days prior to the admission of this patient. He reported a fever 9 days prior to the admission with a maximum temperature of 39.0°C, accompanied by a feeling of chill. After giving himself Radix Bupleuri and Oseltamivir, his temperature returned to normal. He reported a fever of 39.0°C the day at admission.

Laboratory findings

White blood cell count: 4.20×10^9/L, lymphocyte count: 0.81×10^9/L, lymphocyte percentage: 19.2%, C-reactive protein: 5.7 mg/L.

Radiological findings

The radiological findings of the chest CT scan during his hospitalization are as follows: Figs. 2.29–2.32.

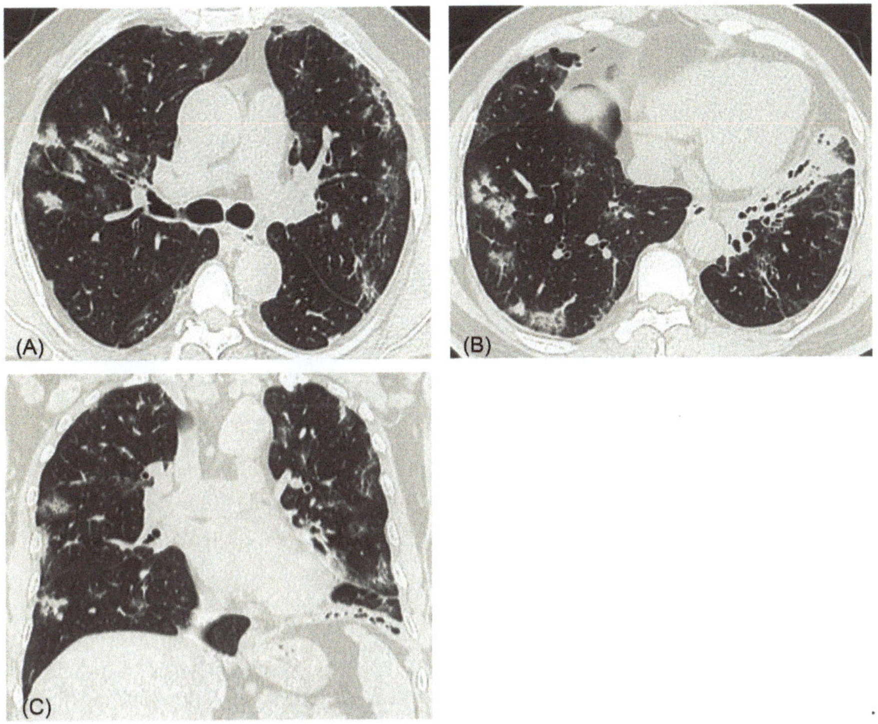

Figure 2.27 Chest computed tomography images 13 days after admission. (A) The ground-glass opacities in the bilateral lungs show slight absorption, with some of them turning into consolidations and parenchymal bands. (B) The consolidations and bronchiectasis in the right lung's lower lobe show mild enlargement. (C) The coronal image shows ground-glass opacities, consolidations and parenchymal bands in the bilateral lungs.

The radiological findings of the chest CT scan after discharge are as follows: Fig. 2.33

Case 19

Medical history

A 20-year-old man reported a history of contact with a patient diagnosed with COVID-19 just 8 days prior to the admission of this patient. He reported a fever on the day of admission with a maximum temperature of 38.0°C, accompanied by cough and expectoration with white sticky sputum.

Laboratory findings

White blood cell count: 6.60×10^9/L, lymphocyte count: 1.59×10^9/L, lymphocyte percentage: 24.1%, C-reactive protein: 6.7 mg/L.

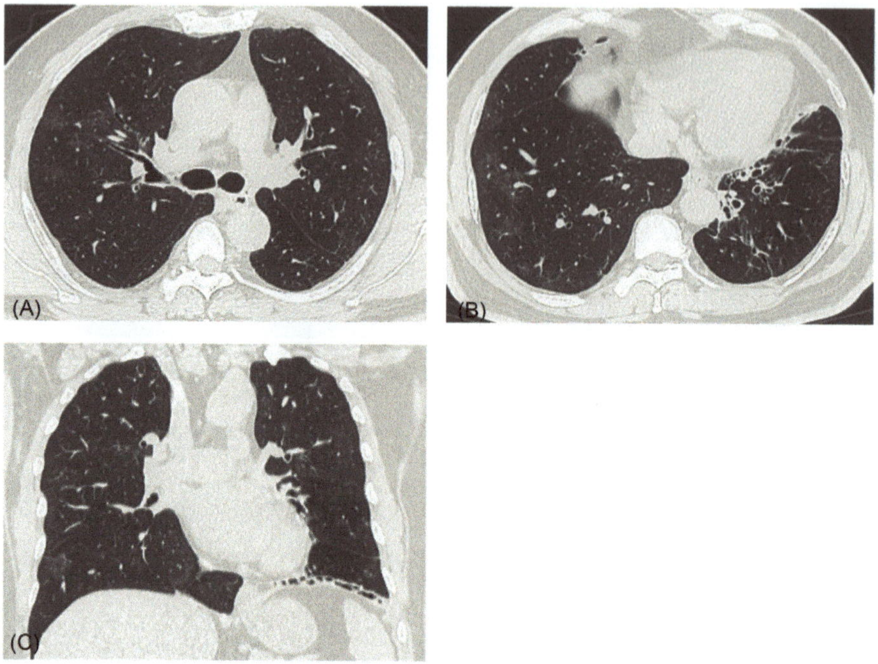

Figure 2.28 Chest computed tomography images 2 months after discharge. (A) A few residual ground-glass opacities in the upper lobes of both lungs is noted. (B) A small amount of ground-glass opacity remains in the lower lobe of the right lung, bronchiectasis in the lower lobe of the left lung, and shrinkage of the surrounding consolidation are noted. (C) The coronal image shows a few residual ground-glass opacities in both lungs.

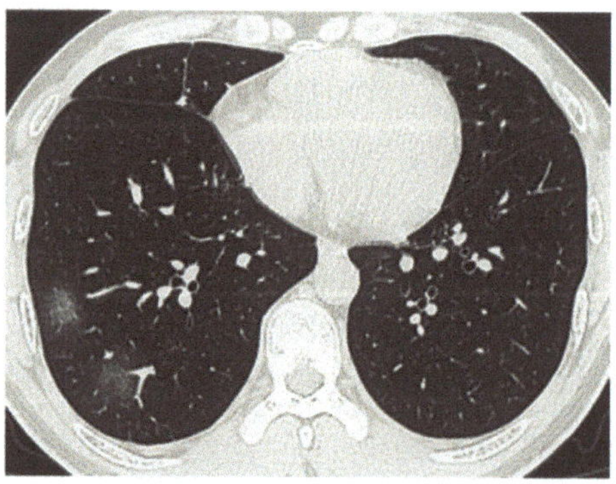

Figure 2.29 Chest computed tomography image on the day of admission. There are multiple ground-glass opacities in the right lung's lower lobe with a peripheral distribution.

Radiological findings

The Radiological Findings of chest CT scans during his hospitalization are as follows: Figs. 2.34–2.37.

The radiological findings of chest CT scan after discharge are as follows: Fig. 2.38.

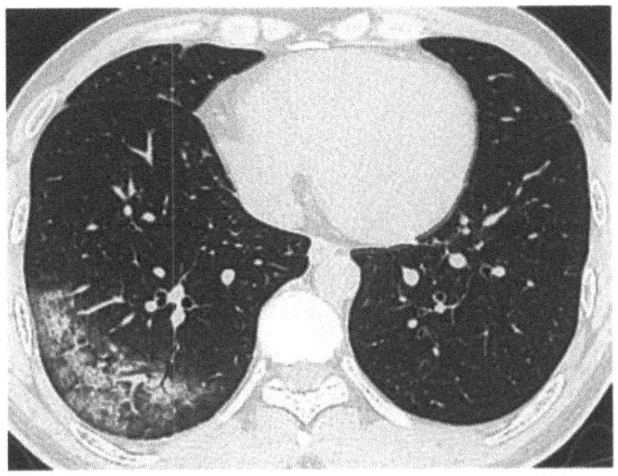

Figure 2.30 Chest computed tomography image 7 days after admission. The ground-glass opacities in the right lung's lower lobe were enlarged and had fused together, with partial consolidation and crazy-paving sign in them.

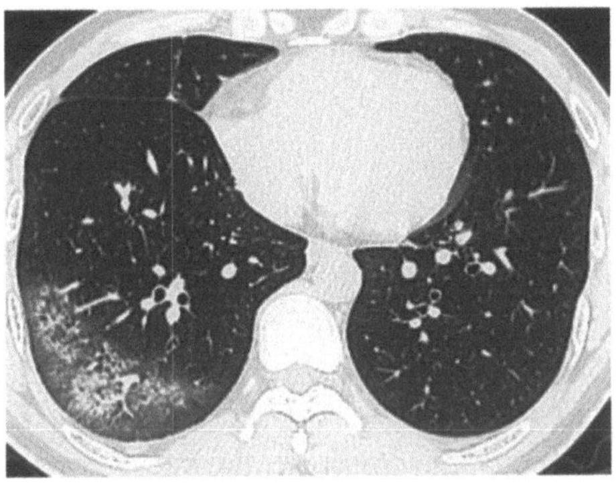

Figure 2.31 Chest computed tomography image 9 days after admission. The lesions in the right lung's lower lobe show absorption, both in size and attenuation.

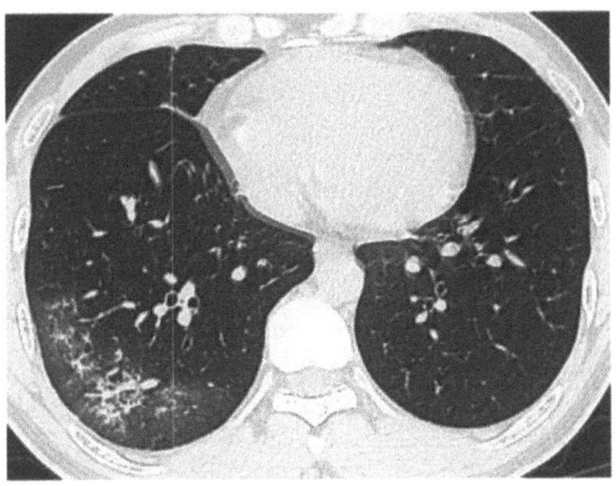

Figure 2.32 Chest computed tomography image 11 days after admission. The lesions in the right lung's lower lobe show great absorption, both in size and attenuation.

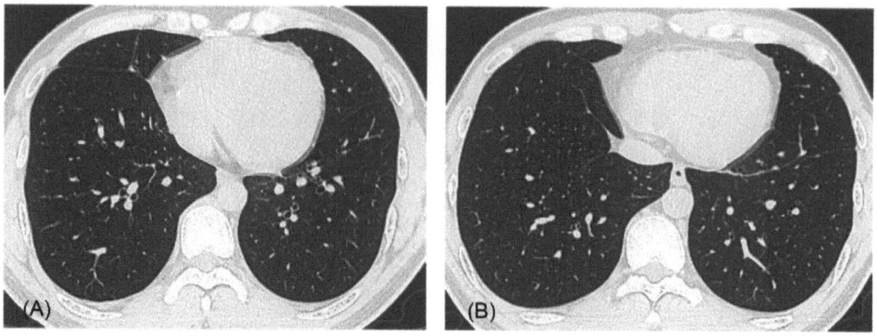

Figure 2.33 Chest computed tomography images 2 months after discharge. (A) The lesions in the right lower lobe are completely absorbed. (B) The lesions in the right lower lobe are completely absorbed.

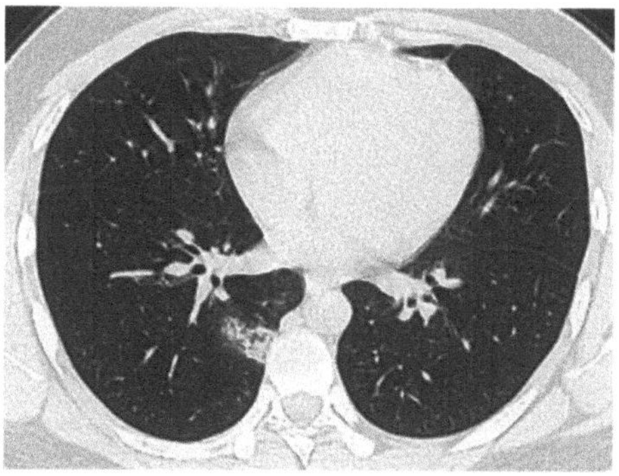

Figure 2.34 Chest computed tomography image on the day of admission. There are ground-glass and consolidative opacities in the medial basal segment of the right lung's lower lobe.

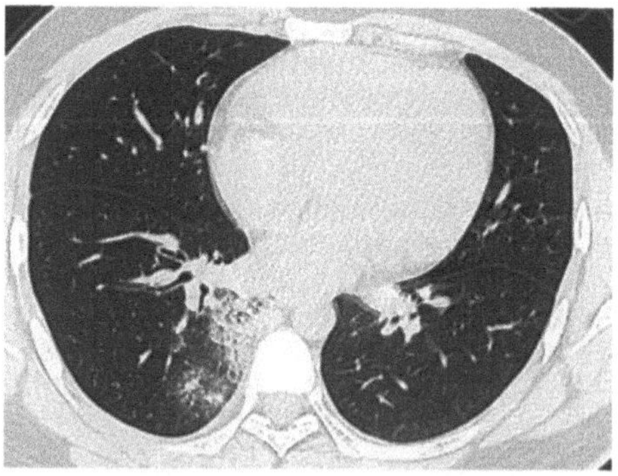

Figure 2.35 Chest computed tomography image 5 days after admission. The ground-glass and consolidative opacities in the medial basal segment of the right lung's lower lobe were enlarged, with an air bronchogram in them.

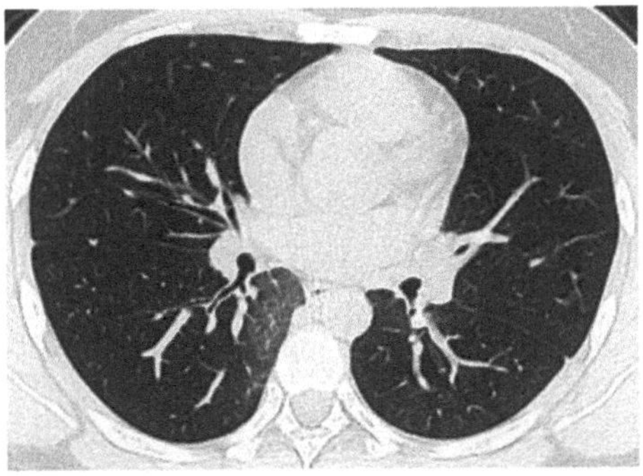

Figure 2.36 Chest computed tomography image 8 days after admission. The lesions in the right lung's lower lobe show obvious absorption, both in size and attenuation.

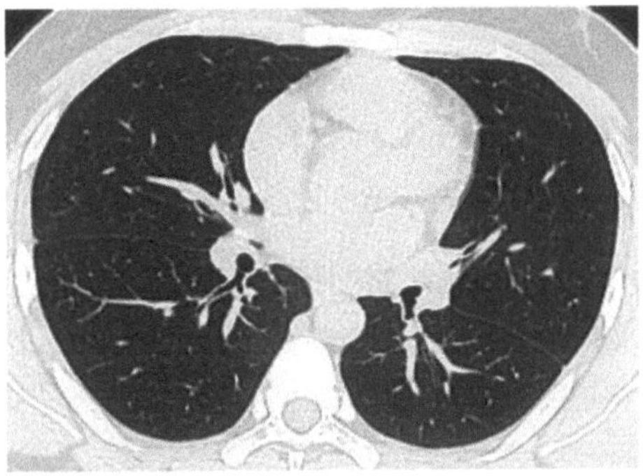

Figure 2.37 Chest computed tomography image 14 days after admission. The lesions in the right lung's lower lobe show almost complete absorption.

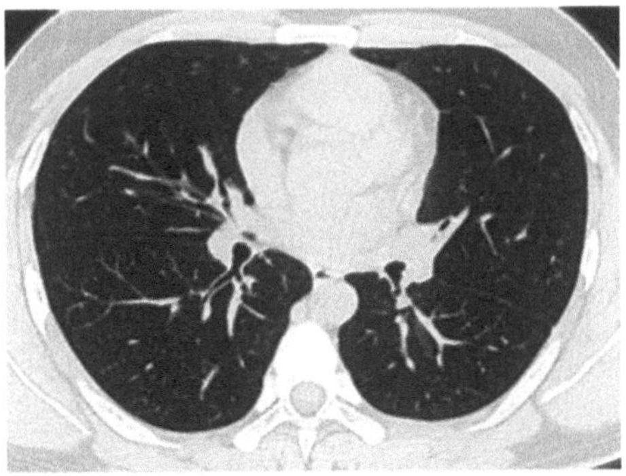

Figure 2.38 Chest computed tomography image 1 month after discharge. Axial chest computed tomography image shows the complete disappearance of lesions in the right lung's lower lobe.

Case 20

Medical history

A 45-year-old woman reported close contact with people who came from the epidemic area just 7 days prior to the admission of this patient. She had a dry throat with occasional chest tightness 2 days prior to the examination by CT.

Laboratory findings

White blood cell count: 3.50×10^9/L, lymphocyte count: 0.76×10^9/L, lymphocyte percentage: 21.4%, C-reactive protein: 4.0 mg/L.

Radiological findings

The radiological findings of the chest CT scan during her hospitalization are as follows: Figs. 2.39–2.43.

Case 21

Medical history

A 57-year-old man lives in the epidemic area, when he left the place just 8 days prior to the admission and reported a fever of 37.7°C just 2 days prior to the admission.

Laboratory findings

White blood cell count: 3.40×10^9/L, lymphocyte count: 1.53×10^9/L, lymphocyte percentage: 45.4%, C-reactive protein: 1.3 mg/L.

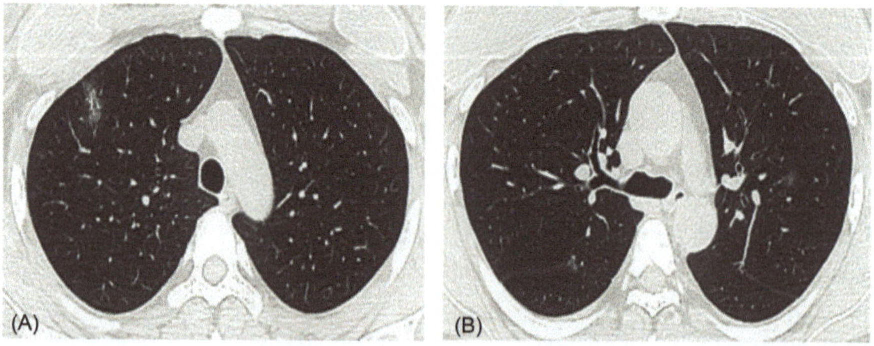

Figure 2.39 Chest computed tomography images on the day of admission. (A) There is a ground-glass opacity with partial consolidation in the right lung's upper lobe; (B) there is a mild ground-glass opacity in the left lung's upper lobe.

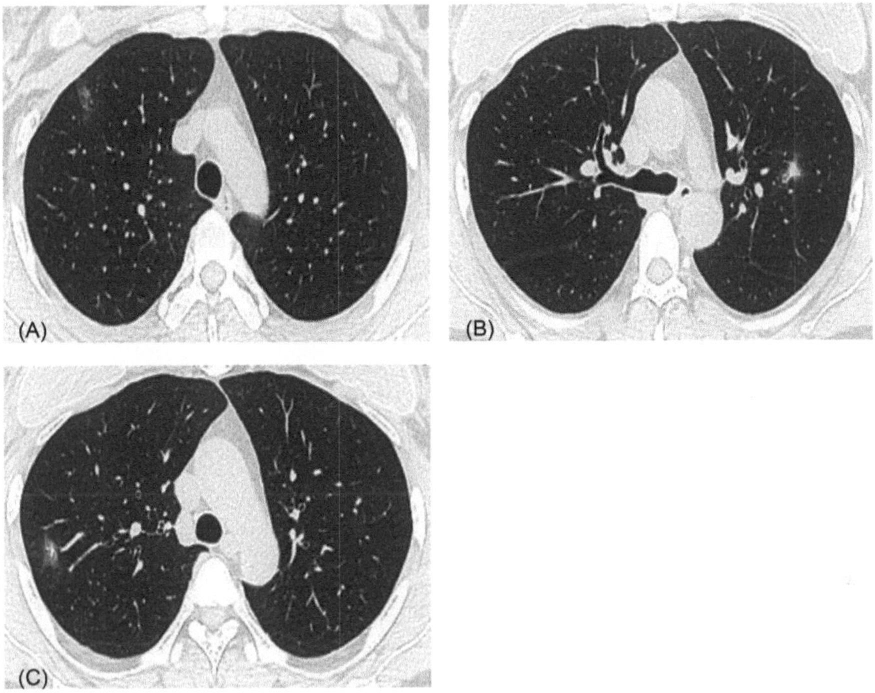

Figure 2.40 Chest computed tomography images 5 days after admission. (A) The lesion in the right lung's upper lobe shows reduction both in area and attenuation, indicating absorption. (B) The ground-glass opacity in the left lung's upper lobe turns into consolidation and enlarges in size. (C) There is a new emerging ground-glass opacity in the posterior segment of the right lung's upper lobe.

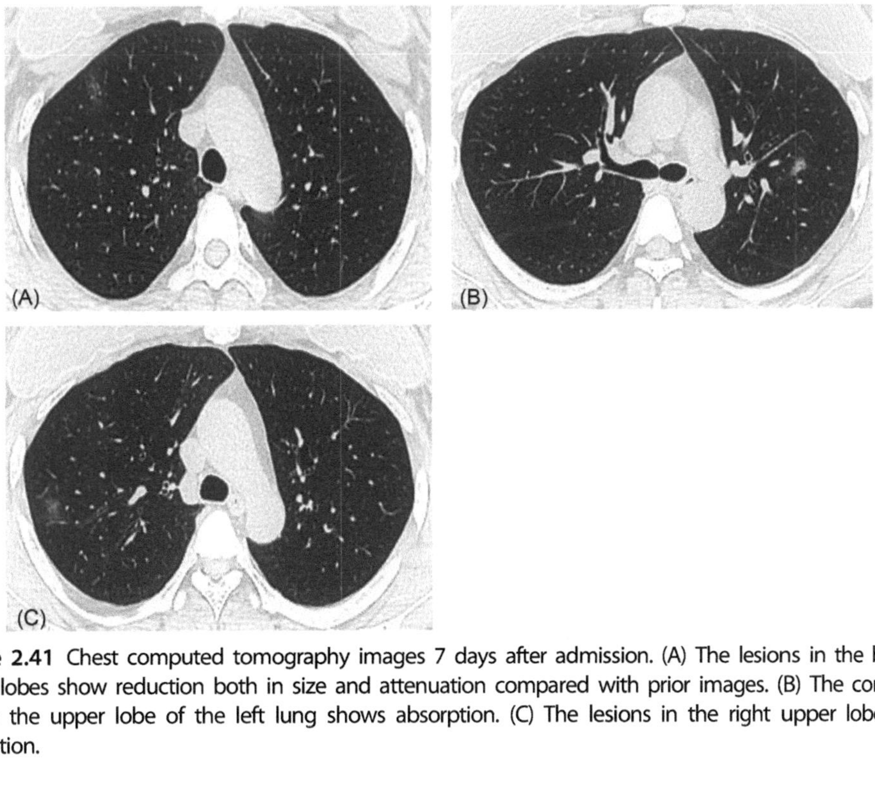

Figure 2.41 Chest computed tomography images 7 days after admission. (A) The lesions in the bilateral upper lobes show reduction both in size and attenuation compared with prior images. (B) The consolidation in the upper lobe of the left lung shows absorption. (C) The lesions in the right upper lobe show absorption.

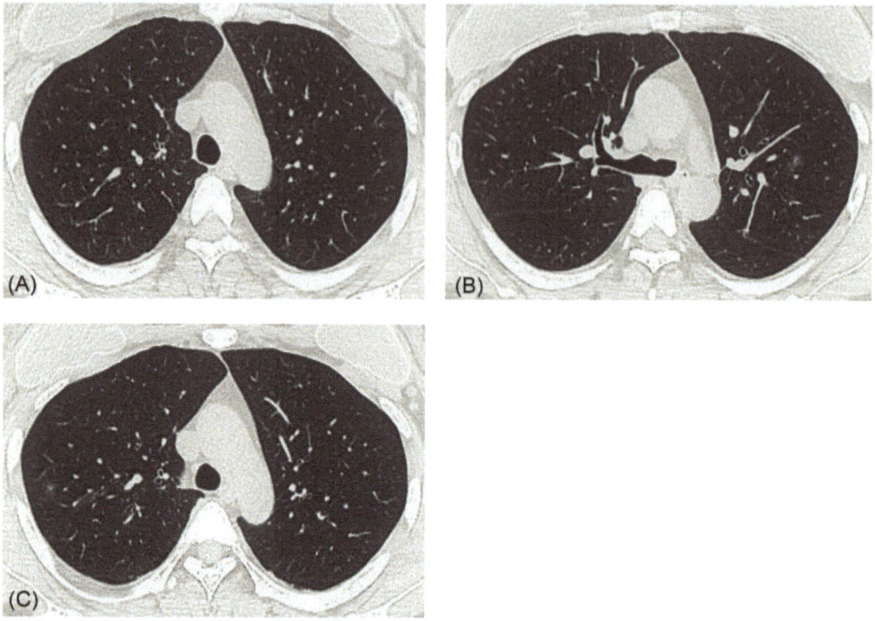

Figure 2.42 Chest computed tomography images 12 days after admission. (A) The lesions in the right upper lobe almost disappear. (B) The lesions in the upper lobe of the left lung show absorption. (C) The lesions in the right upper lobe show absorption.

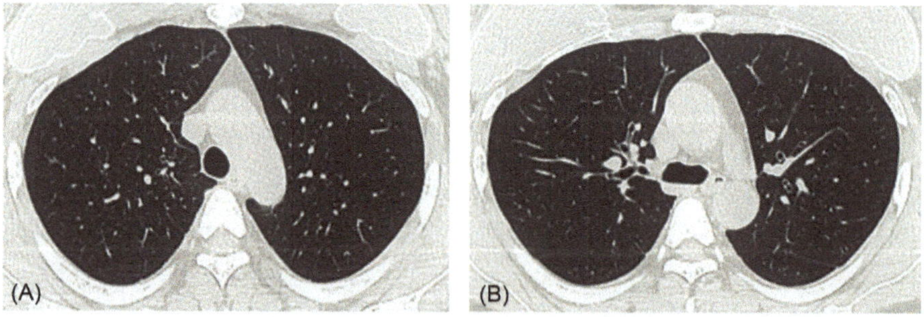

Figure 2.43 Chest computed tomography images 2 months after discharge. (A) Complete absorption of lesions in both lungs is noted. (B) Complete absorption of lesions in both lungs is noted.

Radiological findings

The radiological findings of the chest CT scan during his hospitalization are as follows: Figs. 2.44–2.47.

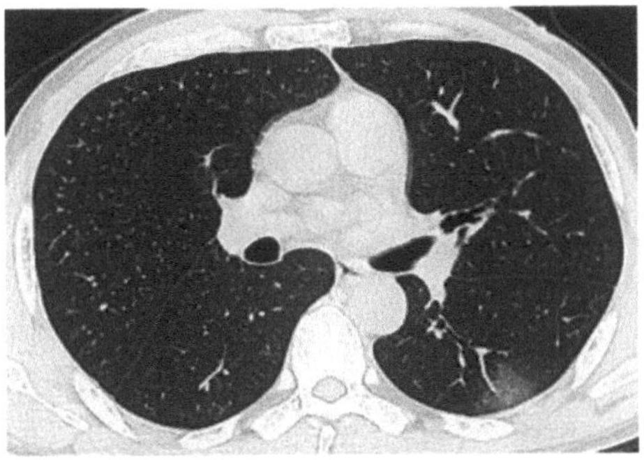

Figure 2.44 Chest computed tomography image on the day of admission. There is a ground-glass opacity in the upper segment of the left lung's lower lobe.

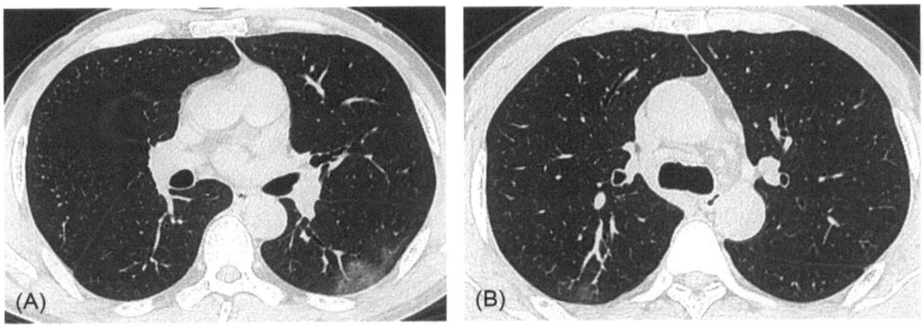

Figure 2.45 Chest computed tomography images 5 days after admission. (A) The ground-glass opacity in the left lung's lower lobe has enlarged, with crazy-paving sign in it; (B) there is a new emerging ground-glass opacity in the posterior segment of the right lung's upper lobe.

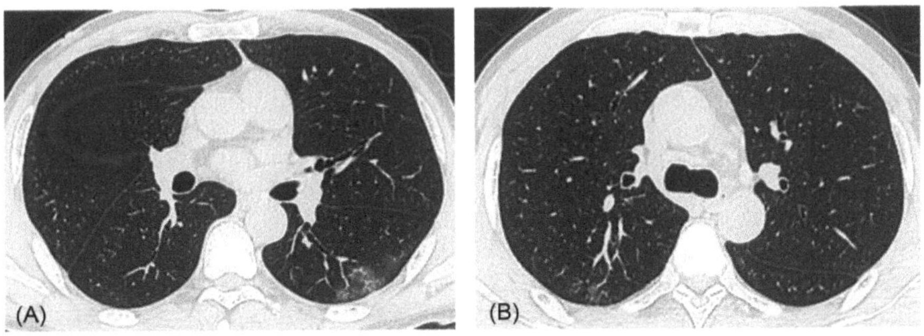

Figure 2.46 Chest computed tomography images 7 days after admission. (A) The left lower lobe lesions show slight absorption. (B) The right lower lobe lesions show slight absorption.

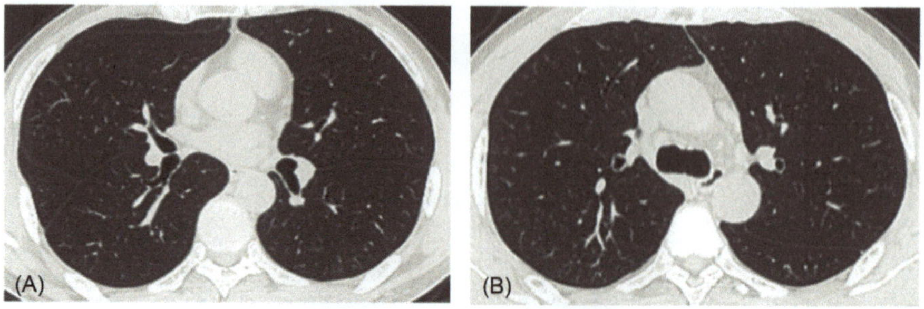

Figure 2.47 Chest computed tomography images 1 month after discharge. (A) The left lower lobe lesions disappear. (B) The right lower lobe lesions disappear.

Case 22

Medical history

A 27-year-old man documented a history of travel to the epidemic area, and he paid a visit again 13 days prior to the admission. He started to have a cough 10 days prior to the admission, and after self-medication, the symptoms went away. He reported intermittent fever 6 days prior to the admission, with a maximum temperature of 37.3°C, accompanied by headache, stuffy nose, sore throat, cough, and expectoration with white sputum.

Laboratory findings

White blood cell count: 3.20×10^9/L, lymphocyte count: 1.57×10^9/L, lymphocyte percentage: 49.3%, C-reactive protein: 19.2 mg/L.

Radiological findings

The radiological findings of the chest CT scan during his hospitalization are as follows: Figs. 2.48 and 2.49.

The radiological findings of chest CT scan after discharge are as follows: Fig. 2.50.

Case 23

Medical history

The patient is a 32-year-old woman who lives in the epidemic area. She left the place just 12 days prior to the admission. She reported a fever 10 days prior to the admission, with a maximum temperature of 39.8°C, accompanied by sore throat, dry cough, and fatigue. She felt chest tightness 1 day prior to the admission.

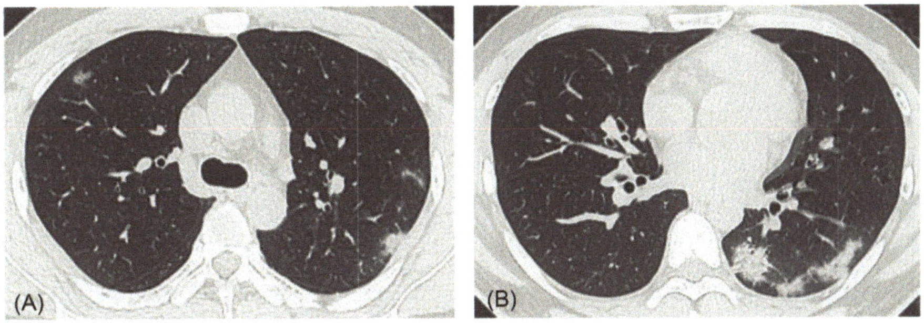

Figure 2.48 Chest computed tomography images 4 days after admission. (A) Multiple subpleural ground-glass opacities and consolidations in the upper lobes of both lungs are noted. (B) Multiple consolidations in the lower lobe of the left lung are noted.

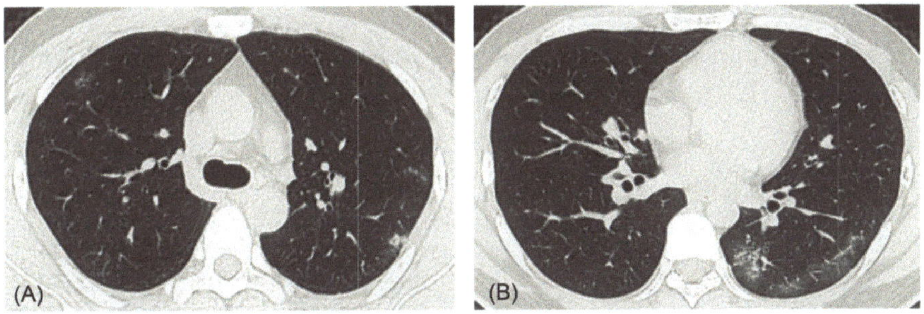

Figure 2.49 Chest computed tomography images 7 days after admission. (A) The lesions in the upper lobes of both lungs show absorption both in size and attenuation. (B) The lesions in the left lower lobe show absorption.

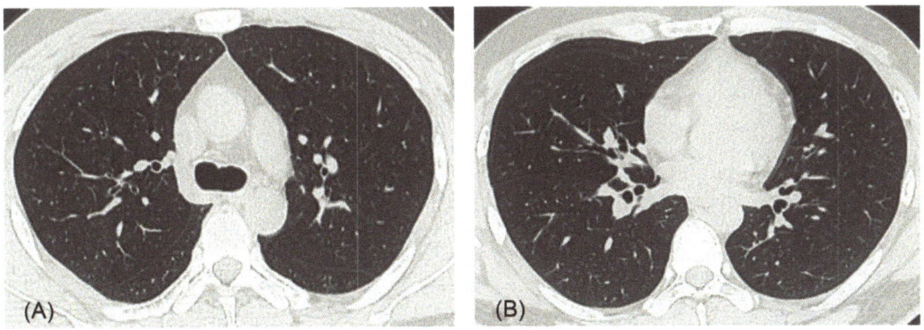

Figure 2.50 Chest computed tomography images 2 months after discharge. (A) Complete absorption of the lesions in the upper lobes of both lungs is noted. (B) The left lower lobe lesions disappear.

Laboratory findings

White blood cell count: $6.50 \times 10^9/L$, lymphocyte count: $1.19 \times 10^9/L$, lymphocyte percentage: 18.5%, C-reactive protein: 127.7 mg/L.

Radiological findings

The radiological findings of the chest CT scan during her hospitalization are as follows: Figs. 2.51–2.53.

Case 24

Medical history

A 25-year-old woman documented a history of contact with people who came from the epidemic area. She reported a fever just 9 days prior to the admission, with a maximum temperature of 39.6°C, accompanied by a feeling of chill, fatigue, poor appetite, cough, and expectoration with white sticky sputum.

Laboratory findings

White blood cell count: $6.20 \times 10^9/L$, lymphocyte count: $0.88 \times 10^9/L$, lymphocyte percentage: 14.2%, C-reactive protein: 29.5 mg/L.

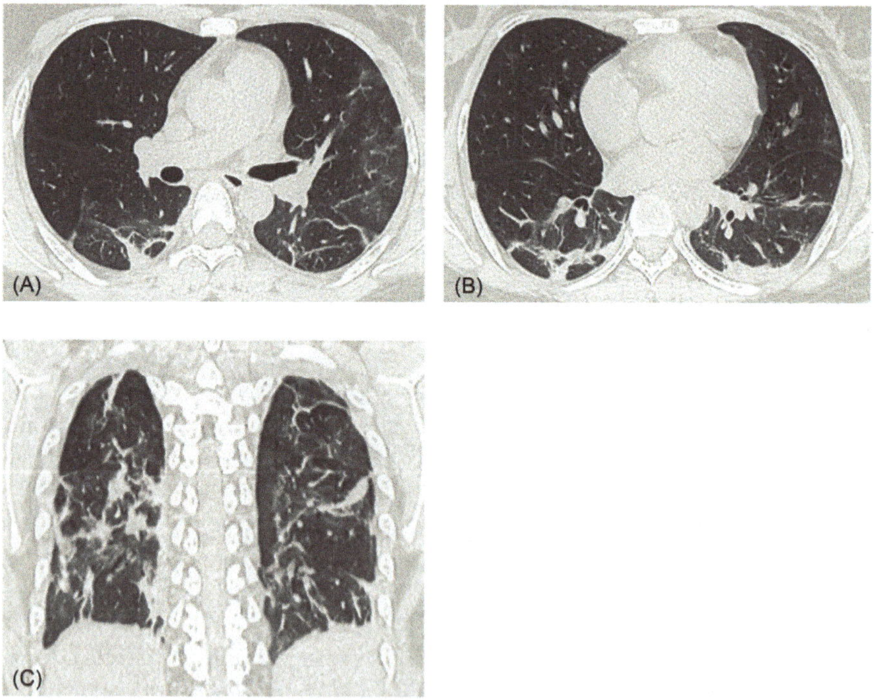

Figure 2.51 Chest computed tomography images 3 days after admission. (A) Multiple ground-glass opacities, consolidations and bands in the upper and lower lobes of the left lung are noted. (B) Multiple ground-glass opacities, consolidations and bands in the lower lobes of both lungs are noted. (C) The coronal image show multiple ground-glass opacities, consolidations and bands in the upper and lower lobes of both lungs, predominant in the lower lobes.

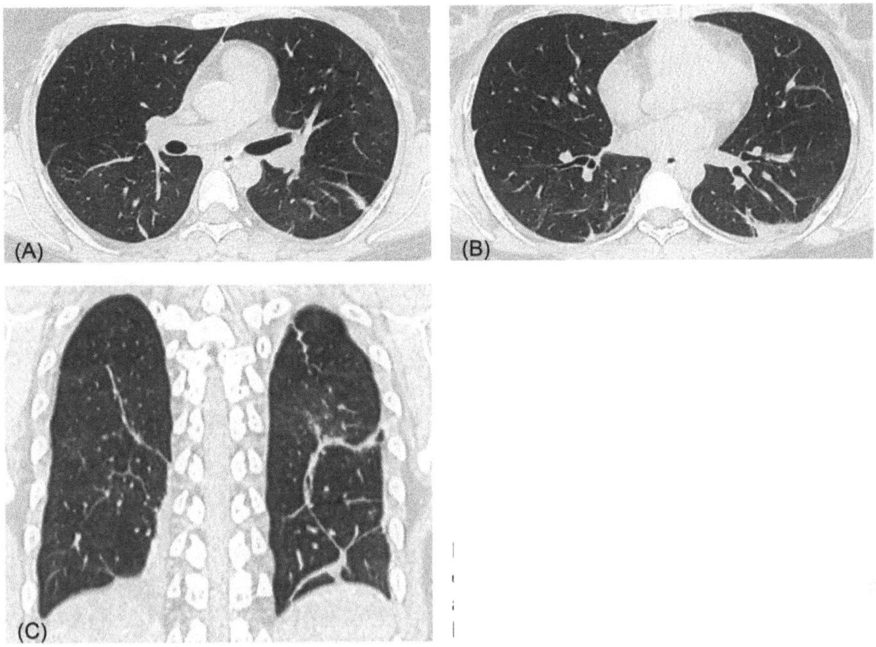

Figure 2.52 Chest computed tomography images 7 days after admission. (A) The lesions in the upper lobe of the left lung and the lower lobes of the two lungs show absorption. (B) The lesions of the lower lobes of the two lungs show absorption. (C) The coronal image show the absorption of lesions in the upper and lower lobes of both lungs.

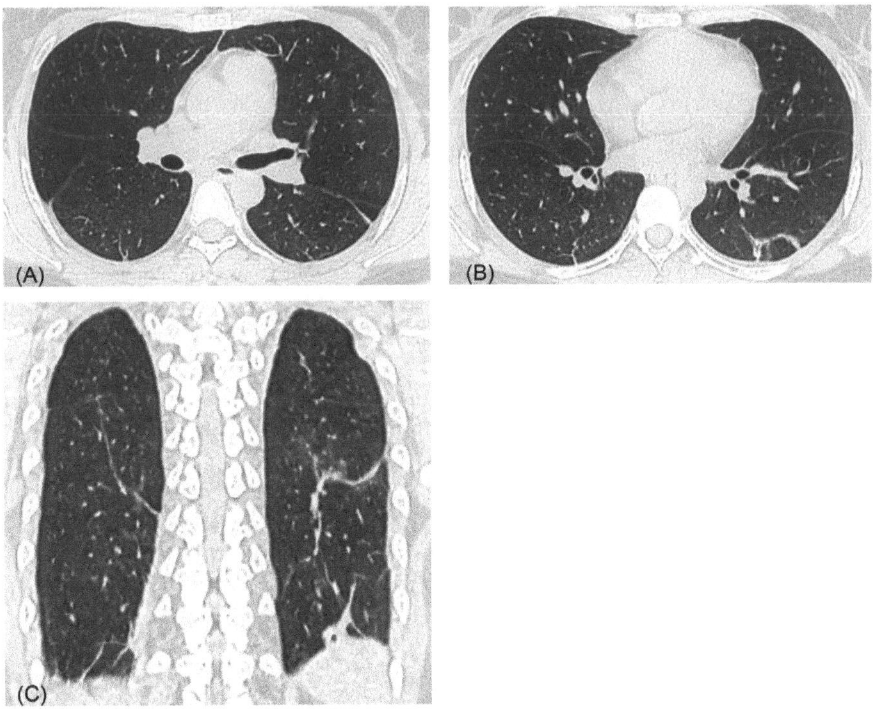

Figure 2.53 admission. (A) The lesions in the upper lobe of the left lung disappear, and the lesions in the lower lobes of the two lungs show continuing absorption. (B) The lesions in the lower lobes of both lungs show continuing absorption. (C) The coronal image shows continuous absorption of lesions in both lungs.

Radiological findings

The radiological findings of the chest CT scan during her hospitalization are as follows: Figs. 2.54 and 2.55.

The radiological findings of chest CT scan after discharge are as follows: Fig. 2.56.

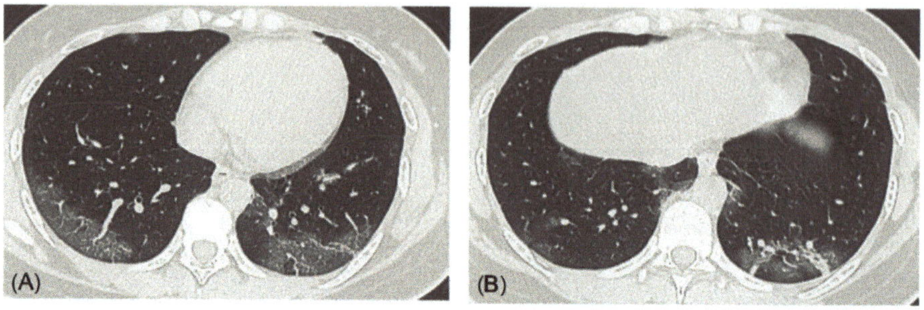

Figure 2.54 Chest computed tomography images on the day of admission. (A) Mild ground-glass opacity in the middle lobe of the right lung, large area of ground-glass opacities in the lower lobes of both lungs are noted, with crazy-paving sign and dilated vessels seen inside. (B) Multiple ground-glass opacities in the lower lobe of the right lung are noted. Ground-glass opacities, consolidations and bands are noted in the lower lobe of the left lung.

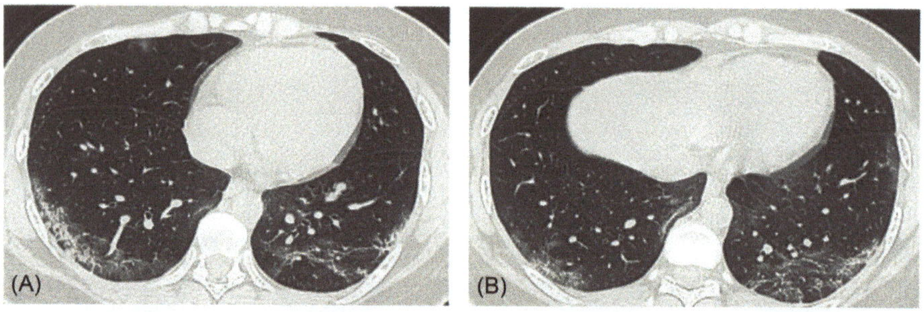

Figure 2.55 Chest computed tomography images 7 days after admission. (A) The ground-glass opacity in the middle lobe of right lung is enlarged. The ground-glass opacities in the bilateral lower lobes are reduced in area, with most of them turning into consolidations and parenchymal bands. (B) The lesions in the bilateral lower lobes enlarge.

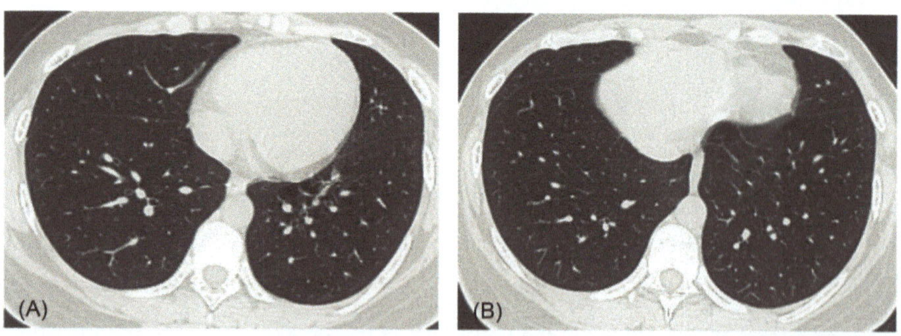

Figure 2.56 Chest computed tomography images 1 month after discharge. (A) The lesions in the right middle lobe and lower lobes of both lungs disappear. (B) The lesions in the lower lobes of both lungs disappear.

Case 25

Medical history

The patient is a 28-year-old man who documented a history of contact with people diagnosed with COVID-19. He reported a fever 5 days prior to the admission, with a maximum temperature of 37.5°C, accompanied by weakness.

Laboratory findings

White blood cell count: 3.30×10^9/L, lymphocyte count: 1.39×10^9/L, lymphocyte percentage: 42.5%, C-reactive protein: 14.6 mg/L.

Radiological findings

The radiological findings of chest CT scan during his hospitalization are as follows: Figs. 2.57–2.60.

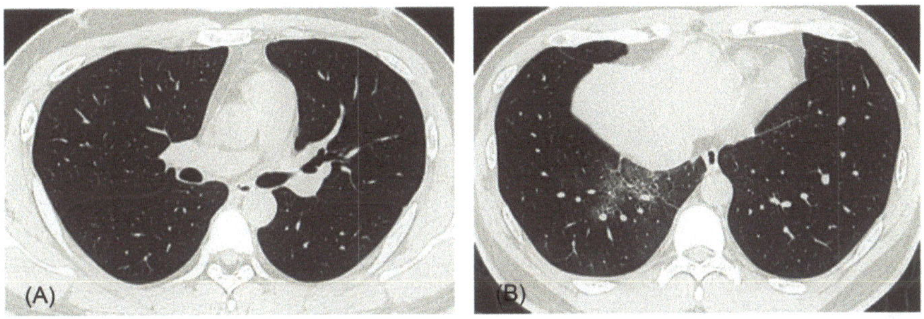

Figure 2.57 Chest computed tomography images on the day of admission. (A) No abnormal findings are noted in the upper and lower lobes of the two lungs. (B) Ground-glass opacities in the lower lobe of the right lung are noted, with thickening of the inter- and intralobular septa (crazy-paving sign) and bronchiolectasis seen inside.

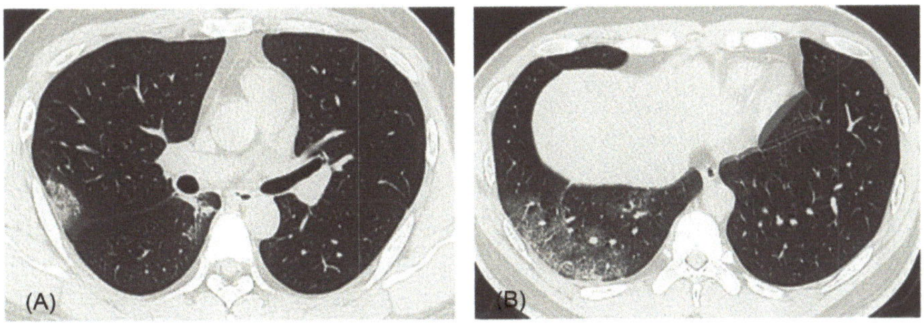

Figure 2.58 Chest computed tomography images 5 days after admission. (A) New ground-glass opacities and consolidations are noted in the upper and lower lobes of the right lung. (B) New ground-glass opacities and consolidations are noted in right lower lobe.

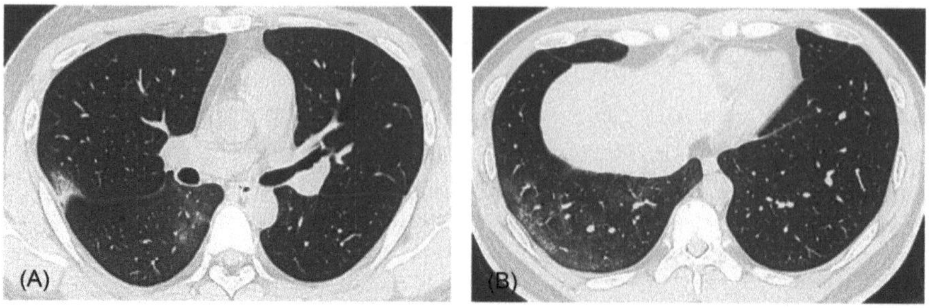

Figure 2.59 Chest computed tomography images 8 days after admission. (A) The lesions in the upper lobe of the right lung show absorption, while the lesions in the lower lobe of the right lung enlarge in area and become lighter in attenuation. (B) The lesions of the right lower lobe show absorption.

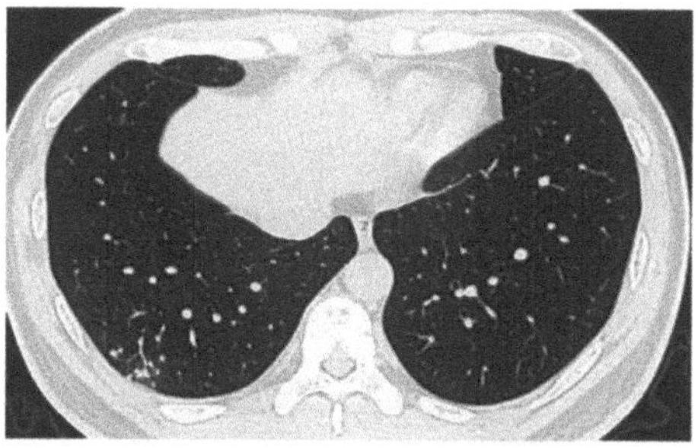

Figure 2.60 Chest computed tomography image 24 days after discharge. Axial chest computed tomography image shows multiple micronodules in the subpleural area of the right lung's lower lobe.

CHAPTER 3

Computed tomography manifestations and prognosis of severe COVID-19 cases

Section 1: Bed-side DR, computed tomography manifestations and prognosis of severe COVID-19 cases

Case 26

Medical history

A 63-year-old man who lives in the epidemic area left the place just 9 days prior to the date of admission. He reported a fever 7 days prior to the date of admission, with a temperature in the range of 37.6°C–39.2°C. He reported cough and expectoration with mild white sticky sputum just 2 days prior to the date of admission with a temperature in the range of 37.8°C–38.9°C.

Laboratory findings

White blood cell count: 5.10×10^9/L, lymphocyte count: 1.23×10^9/L, lymphocyte percentage: 24.0%, C-reactive protein: 12.2 mg/L.

Radiological findings

The radiological findings of the chest CT scans during his hospitalization are as follows: Figs. 3.1–3.4.

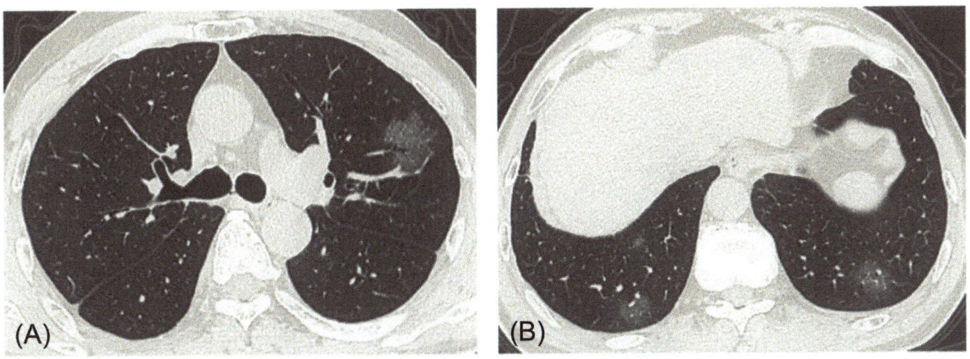

Figure 3.1 Chest computed tomography images on the day of admission. (A) Ground-glass opacities are noted in the left upper lobe. (B) Multiple ground-glass opacities in the lower lobes of both lungs are noted. Thickened intralobular and interlobular septa are noted within the lesions, showing crazy-paving signs.

Image Atlas of COVID-19
DOI: https://doi.org/10.1016/B978-0-443-18493-2.00003-6

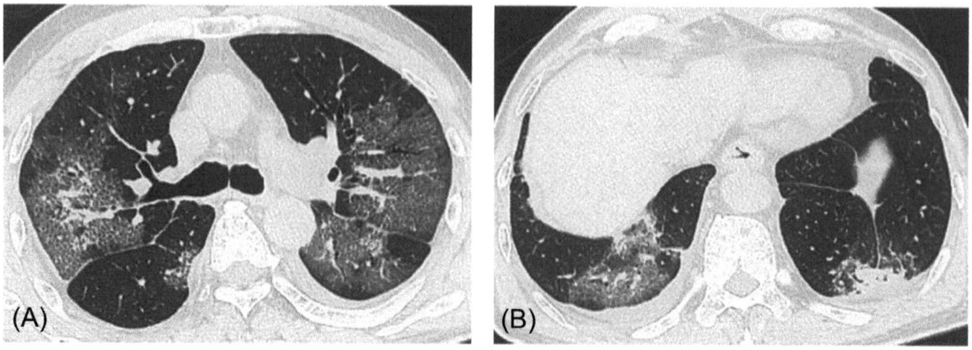

Figure 3.2 Chest computed tomography images 6 days after admission. (A) There are obvious enlargements of the lesions in bilateral lungs, involving more than 1/2 of lung fields. Most of the lesions are ground-glass opacities, with crazy-paving sign, bronchiectasis, and dilation of vessels noticeable in them. (B) Most of the lesions in the left lung's lower lobe are consolidations. There is a mild amount of pleural effusion in bilateral thoracic cavities.

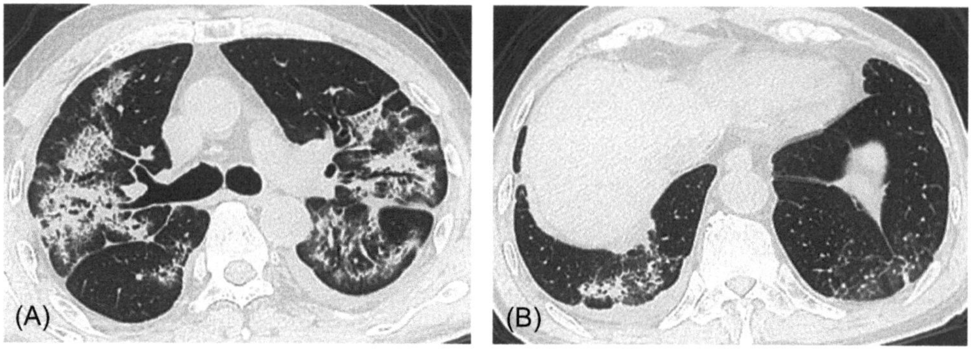

Figure 3.3 Chest computed tomography images 14 days after admission. (A) There is an obvious reduction in the area of ground-glass opacities in both lungs. Most of them turn into consolidations and bands, with distorted structure and bronchiectasis seen. (B) The lesions in the lower lobe of the left lung show absorption, with ground-glass opacities and bands. The lesions in the right lower lobe show reduction in area.

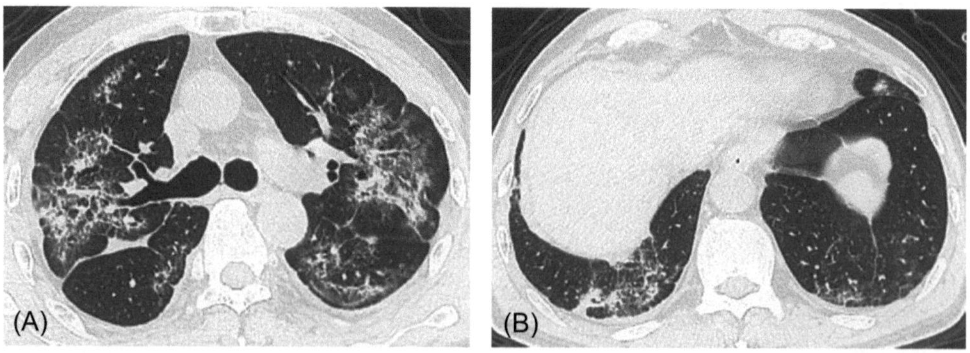

Figure 3.4 Chest computed tomography images 25 days after admission. (A) The lesions in the upper and lower lobes of both lungs are further absorbed, with a reduction in both area and attenuation, however, the distorted structure and parenchymal bands are still seen. (B) The lesions in the lower lobes of both lungs are further absorbed.

Case 27

Medical history

A 56-year-old man documented a travel history to the epidemic area. He reported a fever 4 days prior to the date of admission, with a maximum temperature of 38.0°C, accompanied by weakness. He started to have cough and chest tightness 1 day prior to the date of admission.

Laboratory findings

White blood cell count: $6.00 \times 10^9/L$, lymphocyte count: $0.78 \times 10^9/L$, lymphocyte percentage: 13.0%, C-reactive protein: 140.4 mg/L.

Radiological findings

The radiological findings of chest CT scans during his hospitalization are as follows: Figs. 3.5–3.8.

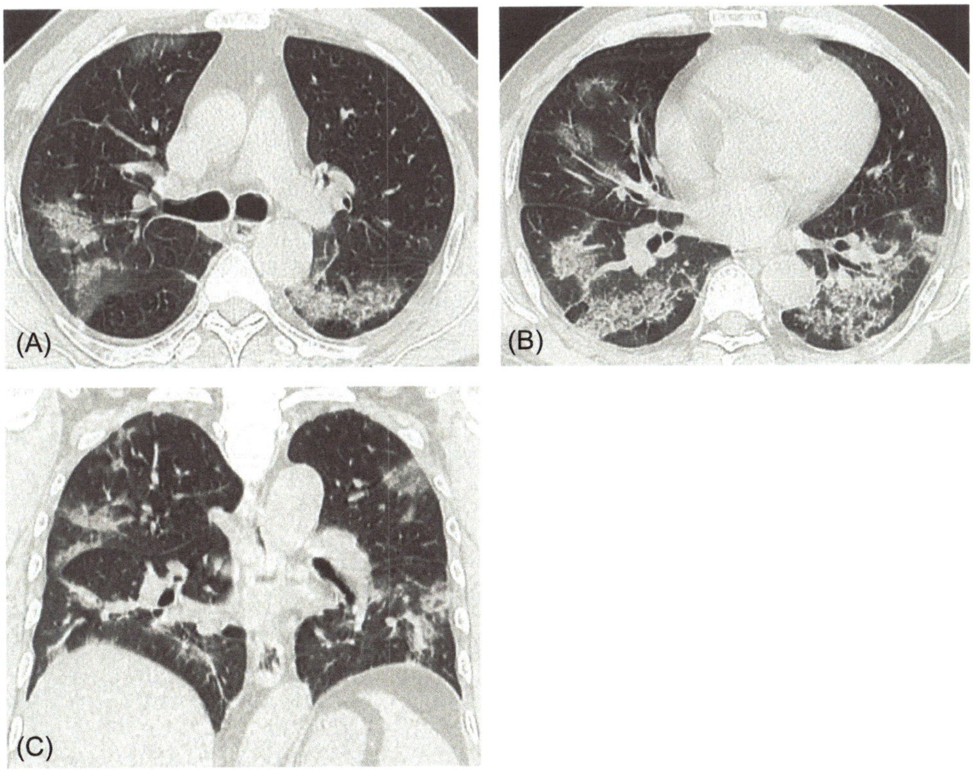

Figure 3.5 Chest computed tomography images on the day of admission. (A) Multiple ground-glass opacities and consolidations are noted in the left lower lobe and right upper lobe, showing crazy-paving signs. (B) Multiple ground-glass opacities and consolidations are noted in the upper and lower lobes of both lungs, showing crazy paving signs. (C) The coronal image shows multiple ground-glass opacities and consolidations in the upper and lower lobes of both lungs.

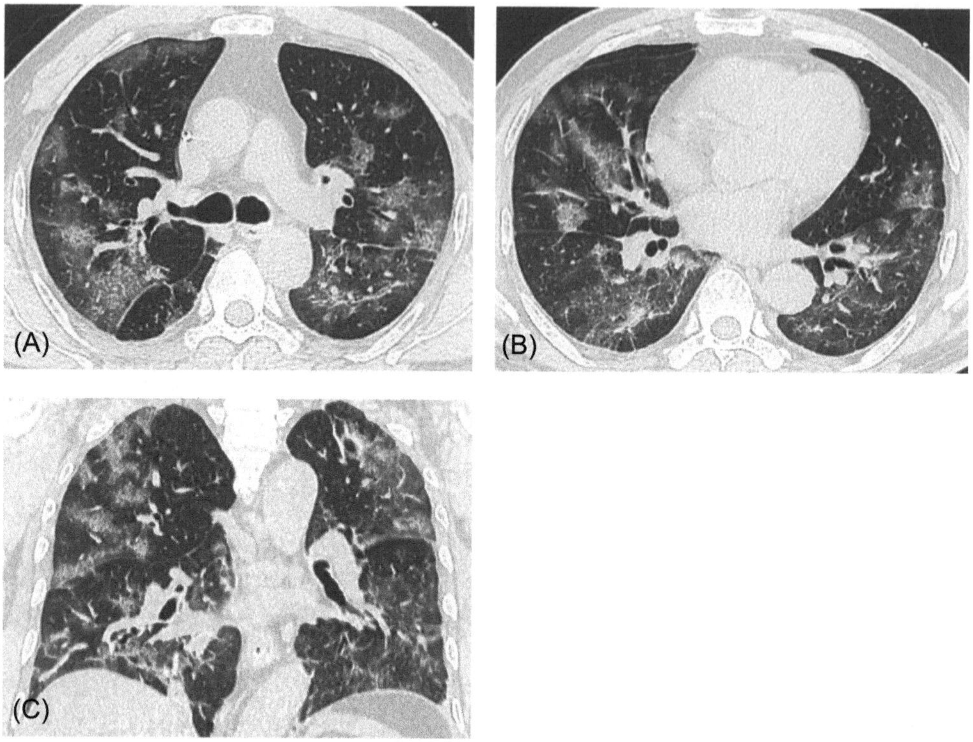

Figure 3.6 Chest computed tomography images 5 days after admission. (A) The area of lesions enlarges in both lungs, with more than 1/2 of the lung fields involved, while the attenuation of the lesions reduces. (B) Multiple new ground-glass opacities are noted in the upper and lower lobes of both lungs, the original lesions are partially absorbed. (C) The coronal image shows multiple new ground-glass opacities in the upper and lower lobes of both lungs, the original lesions are partially absorbed.

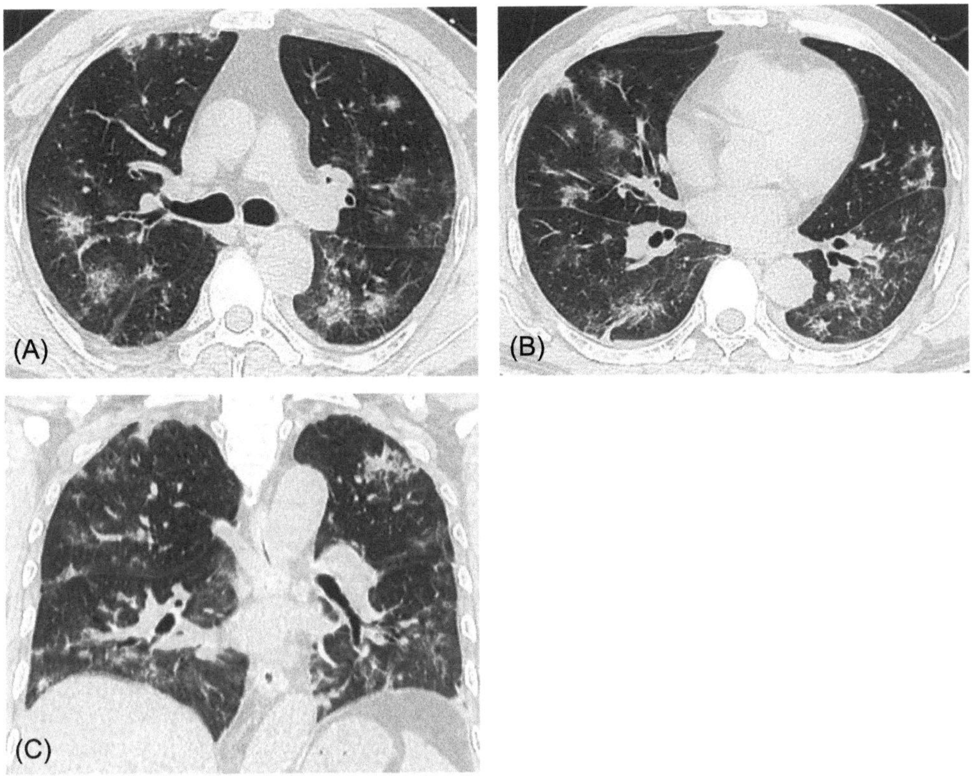

Figure 3.7 Chest computed tomography images 9 days after admission. (A) The lesions in the upper and lower lobes of both lungs are partially absorbed, with reduction in area and attenuation. (B) The lesions in the upper and lower lobes of both lungs are partially absorbed. (C) The coronal image shows partial absorption of lesions in the upper and lower lobes of both lungs.

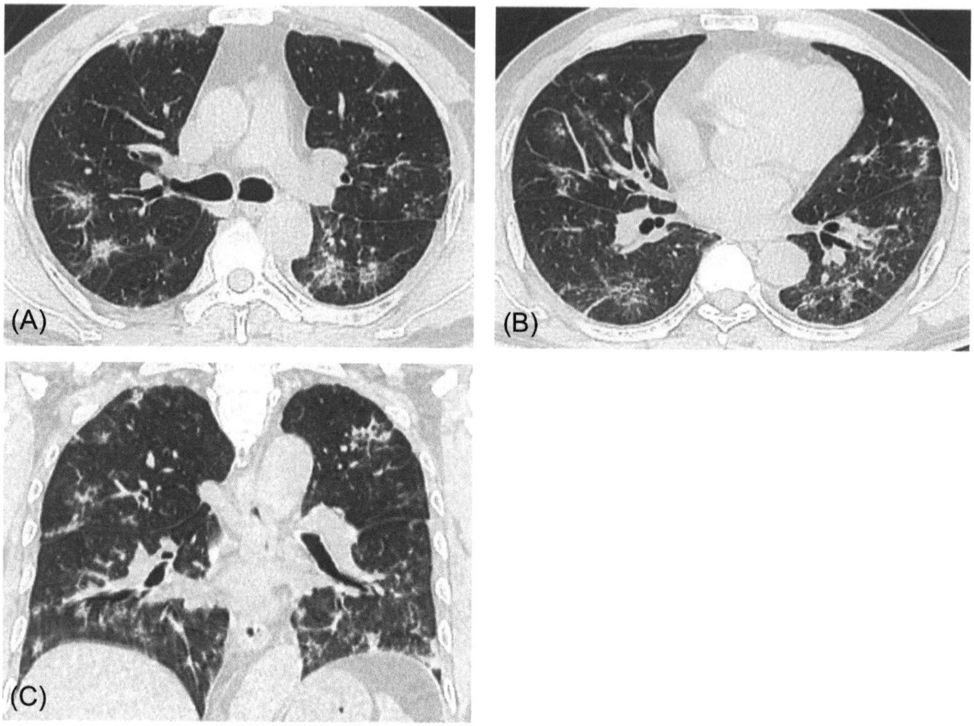

Figure 3.8 Chest computed tomography images 16 days after admission. (A) The lesions in the upper and lower lobes of both lungs show continuing absorption. (B) The lesions in the upper and lower lobes of both lungs are partially absorbed, with bronchiectasis and parenchymal bands seen inside. (C) The lesions in the upper and lower lobes of both lungs are partially absorbed.

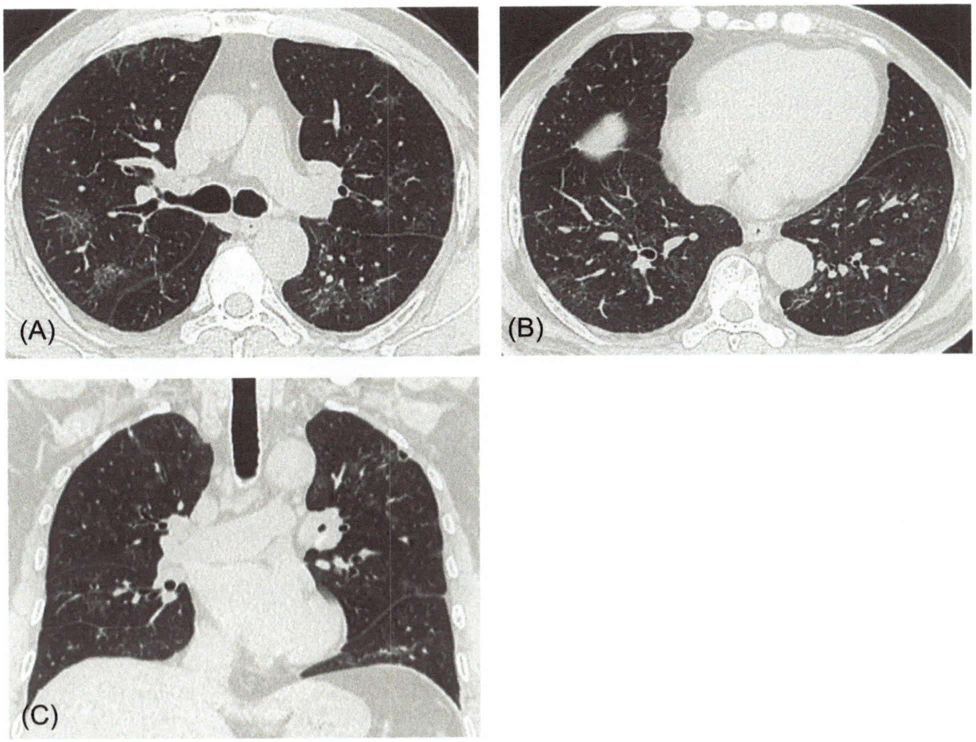

Figure 3.9 Chest computed tomography images one and a half months after discharge. (A) The lesions in the upper and lower lobes of both lungs are almost completely absorbed, with reduction in size and attenuation. (B) The lesions in the upper and lower lobes of both lungs are almost completely absorbed. (C) The coronal image shows that the lesions in the upper and lower lobes of both lungs are almost completely absorbed.

The radiological findings of the chest CT scans after discharge are as follows: Fig. 3.9.

Case 28

Medical history

A 32-year-old man lives in the epidemic area and left the place 10 days prior to the date of admission. He reported fever and chills 8 days prior to the date of admission, with a maximum temperature of 38.0°C, accompanied by expectoration after his getting up in the morning.

Laboratory findings

White blood cell count: 9.60×10^9/L, lymphocyte count: 0.38×10^9/L, lymphocyte percentage: 4.0%, C-reactive protein: 33.0 mg/L.

Radiological findings

The radiological findings of bed-side DR and chest CT scans during his hospitalization are as follows: Figs. 3.10–3.14.

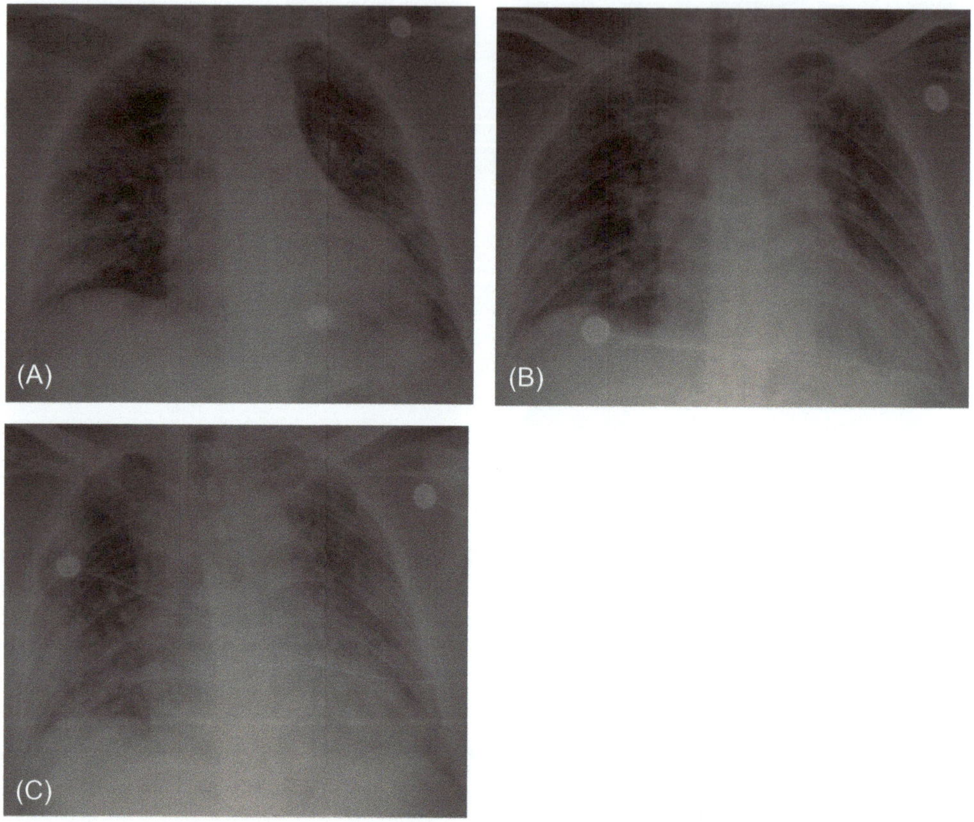

Figure 3.10 (A) Bedside DR 2 days after admission, blurred lung marking and obscure patchy shadows in both lungs are noted. (B) Bedside DR 4 days after admission, the lesions increase. (C) Bedside DR 8 days after admission, the lesions increase gradually.

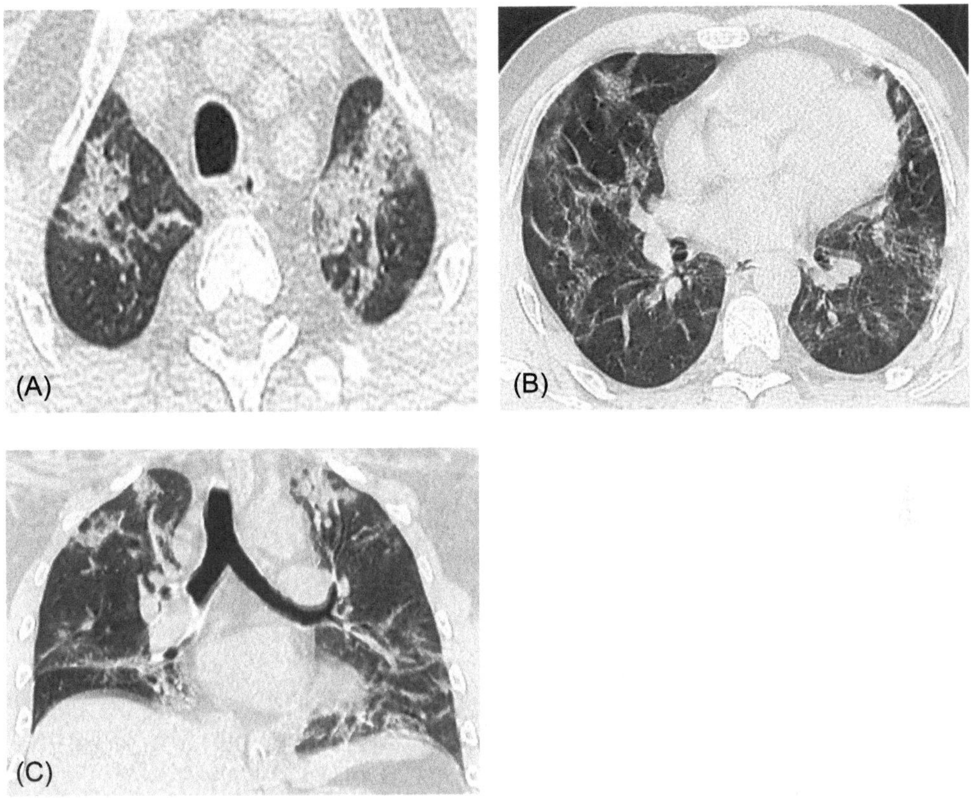

Figure 3.11 Computed tomography images 11 days after admission. (A) Consolidations and bands are noted in the upper lobe of both lungs. (B) Multiple ground-glass opacities and bands are noted in the upper and lower lobes of both lungs, with a distorted structure. (C) The coronal image shows consolidations, ground-glass opacities in the upper lobe of both lungs, multiple fibrotic bands and ground-glass opacities in the lower lobe of both lungs.

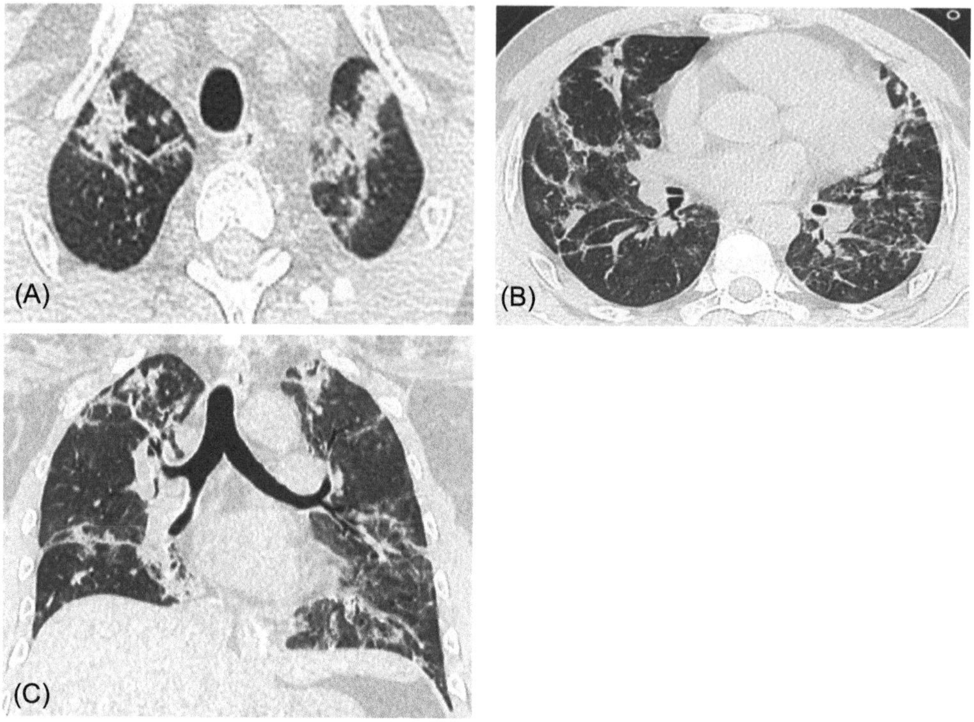

Figure 3.12 Computed tomography images 14 days after admission. (A) The upper lobe lesions in both lungs are partially absorbed, with reduced extent and increased attenuation. (B) The lesions in the upper and lower lobes of both lungs are partially absorbed. (C) The coronal image shows the upper lobe lesions in both lungs are partially absorbed.

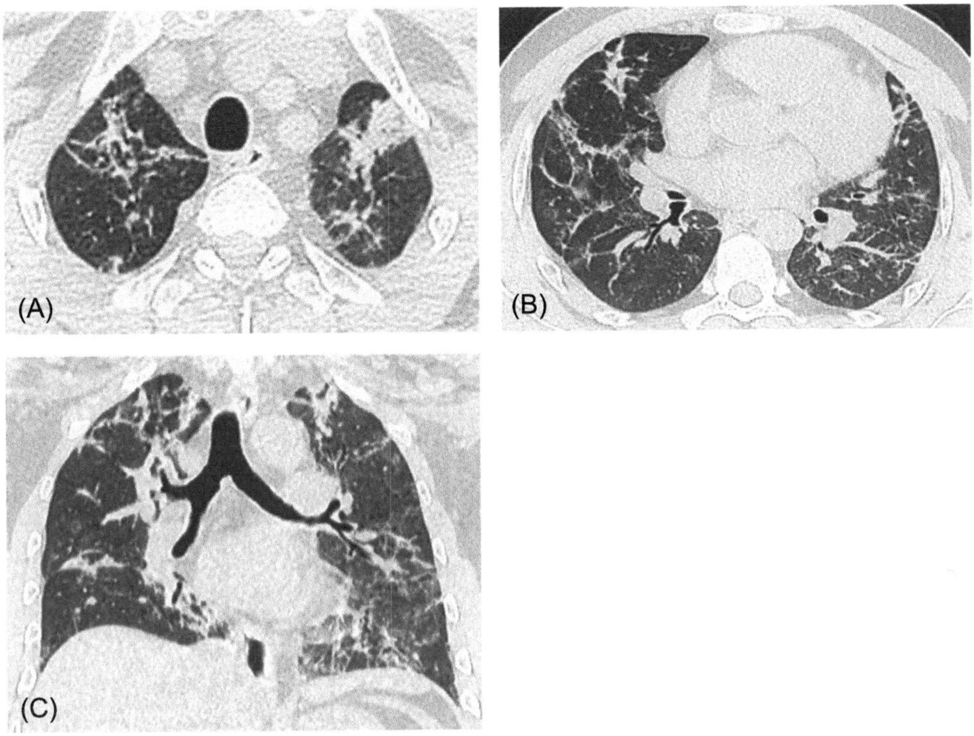

Figure 3.13 Computed tomography images 18 days after admission. (A) The upper lobe lesions of both lungs are partially absorbed. (B) The upper lobe lesions in both lungs are partially absorbed. (C) The coronal image shows the lower lobe lesions in both lungs are partially absorbed compared with prior images.

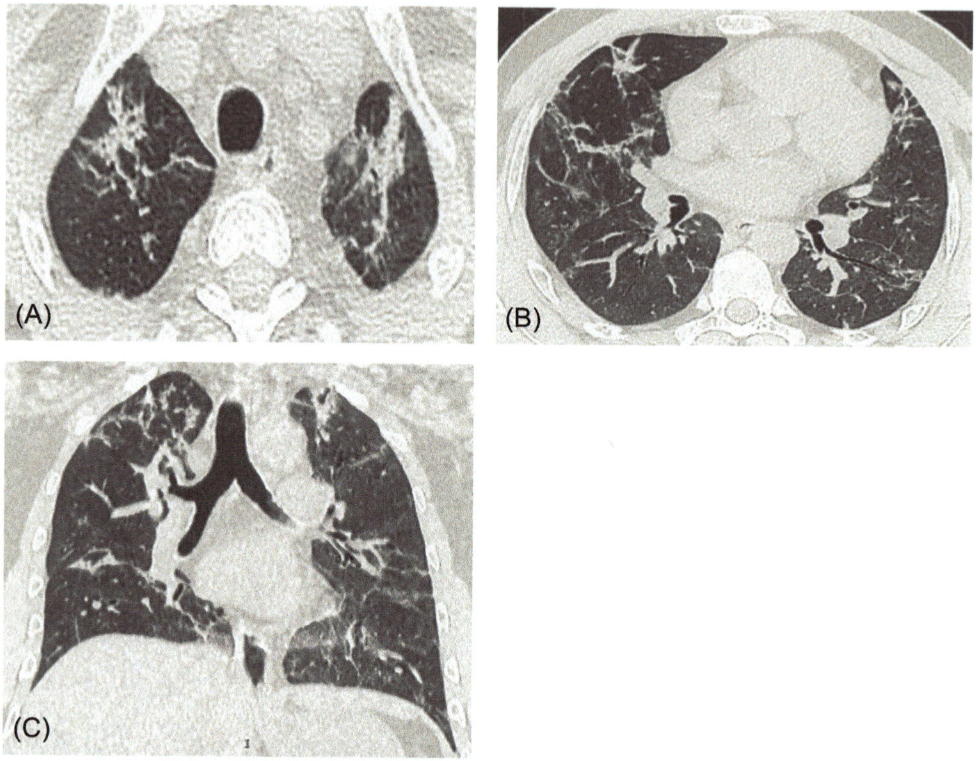

Figure 3.14 Computed tomography images 23 days after admission. (A) The lesions in the upper lobes of both lungs are further absorbed. (B) The lesions in the upper and lower lobes of both lungs are further absorbed. (C) The coronal image shows further absorption of the lesions in the upper and lower lobes of both lungs.

Case 29

Medical history

A 51-year-old woman documented a history of contact with people who were diagnosed with COVID-19. She reported fever 4 days prior to the admission, with a maximum temperature of 38.3°C, accompanied by weakness and poor appetite. She responded poorly to traditional Chinese antiviral medicine.

Laboratory findings

White blood cell count: 4.30×10^9/L, lymphocyte count: 1.13×10^9/L, lymphocyte percentage: 26.3%, C-reactive protein: 23.1 mg/L.

Radiological findings

The radiological findings of chest CT scans during her hospitalization are as follows: Figs. 3.15–3.18.

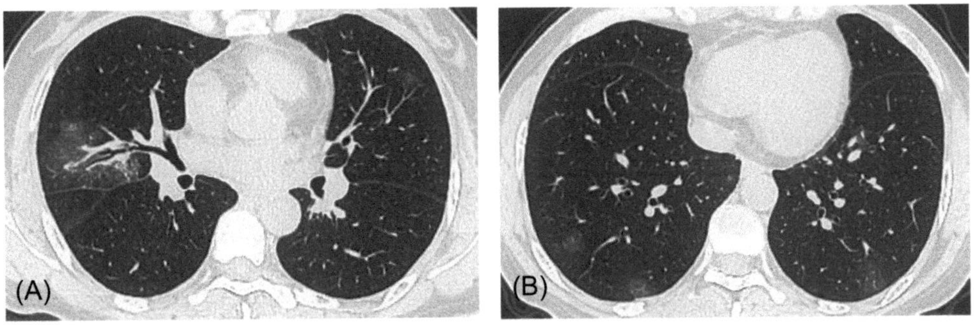

Figure 3.15 Chest computed tomography images on the day of admission. (A) There are ground-glass opacities in both lungs, mainly in the right lung. (B) Ground-glass opacities in the lower lobes of both lungs are noted.

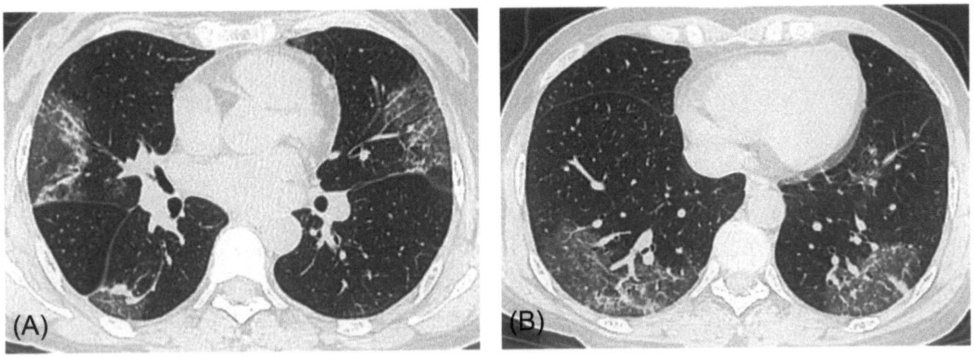

Figure 3.16 Chest computed tomography images 10 days after admission. (A) The lesions in both lungs enlarge significantly, with consolidations, thickening of the inter- and intralobular septa and bands in the ground-glass opacities. The dilation of vessels in the lesions is more obvious than before. (B) The enlargements of ground-glass opacities and consolidations in the lower lobes of both lungs are noted.

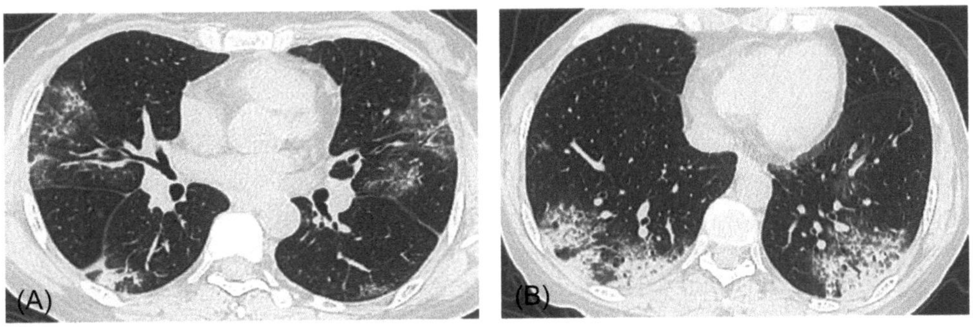

Figure 3.17 Chest computed tomography images 13 days after admission. (A) The lesions in the bilateral upper lungs show absorption, with a reduction in area, while the structure is distorted in the lesions, with bands and bronchiectasis in them. (B) The attenuation of the bilateral lower lesions increases, with part of them turning into consolidations.

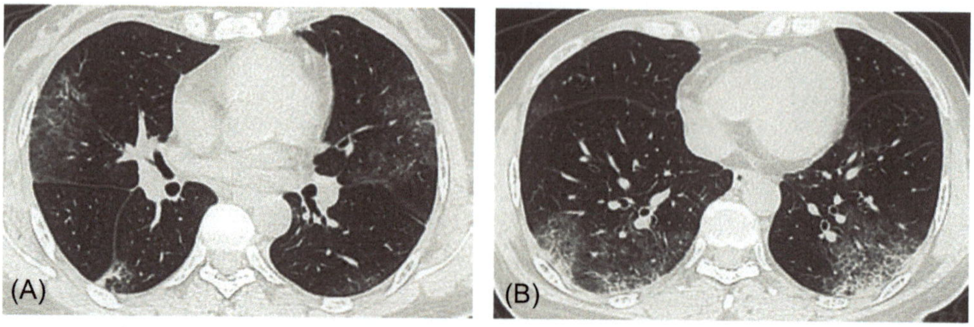

Figure 3.18 Chest computed tomography images 16 days after admission. (A) The lesions in the upper and lower lobes of both lungs are markedly absorbed, with reduced extent and attenuation. (B) The lesions in the lower lobes of both lungs are markedly absorbed, with reduced extent and attenuation.

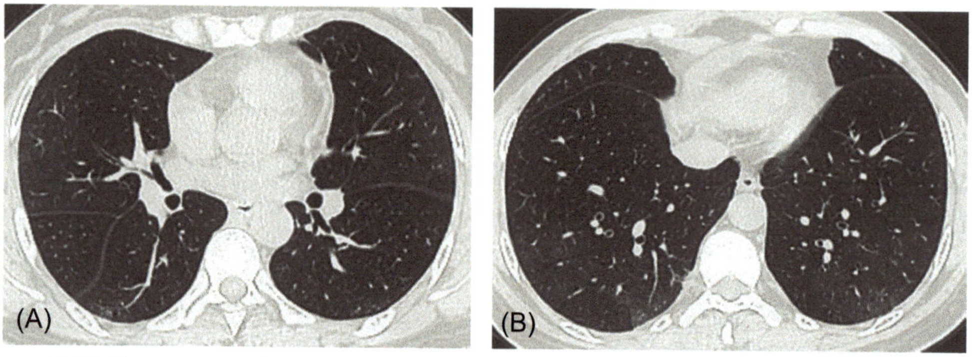

Figure 3.19 Chest computed tomography images one and a half months after discharge. (A) The lesions in the upper and lower lobes of both lungs are almost completely absorbed. (B) The lesions in the lower lobes of both lungs are almost completely absorbed, with a few ground-glass opacities remaining.

The radiological findings of the chest CT scans after discharge are as follows: Fig. 3.19.

Case 30

Medical history

A 78-year-old woman documented a contact history with people who came from the epidemic area. She reported a fever about 3 weeks prior to the admission date, with a maximum temperature of 38.0°C, accompanied by cough, expectoration with white sticky sputum, fatigue, and palpitation. The throat swab nucleic acid test was positive 18 days prior to the date of admission. The body temperature returned to normal after receiving antiviral, antibiotic, and traditional Chinese medicine treatment, while cough and expectoration did not go away. She reported chest tightness and palpitation 2 days prior to the date of admission.

Laboratory findings

White blood cell count: 6.50×10^9/L, lymphocyte count: 1.22×10^9/L, lymphocyte percentage: 18.8%, C-reactive protein: 27.9 mg/L.

Radiological findings

The radiological findings of chest CT scans during her hospitalization are as follows: Figs. 3.20 and 3.21.

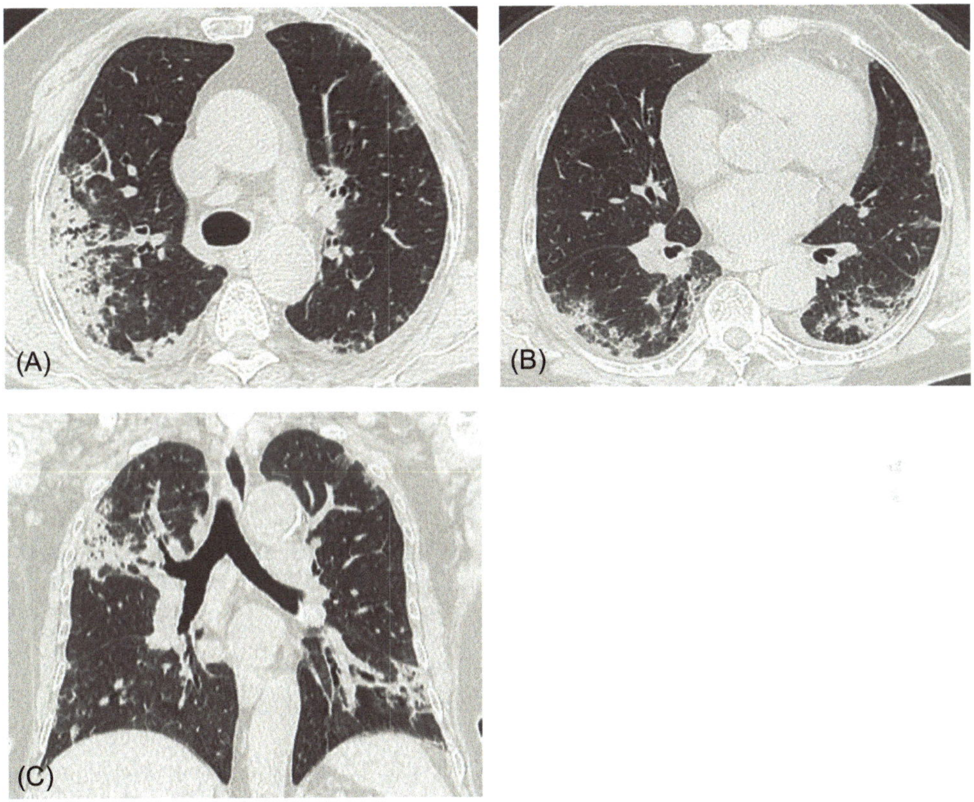

Figure 3.20 Chest computed tomography images on the day of admission. (A) Multiple ground-glass opacities, consolidations and bands with bronchiectasis in both lungs are noted, mainly distributed in the peripheral subpleural area. (B) Multiple ground-glass opacities, consolidations and bands with bronchiectasis in both lungs are noted. (C) The coronal image shows consolidations and ground-glass opacities.

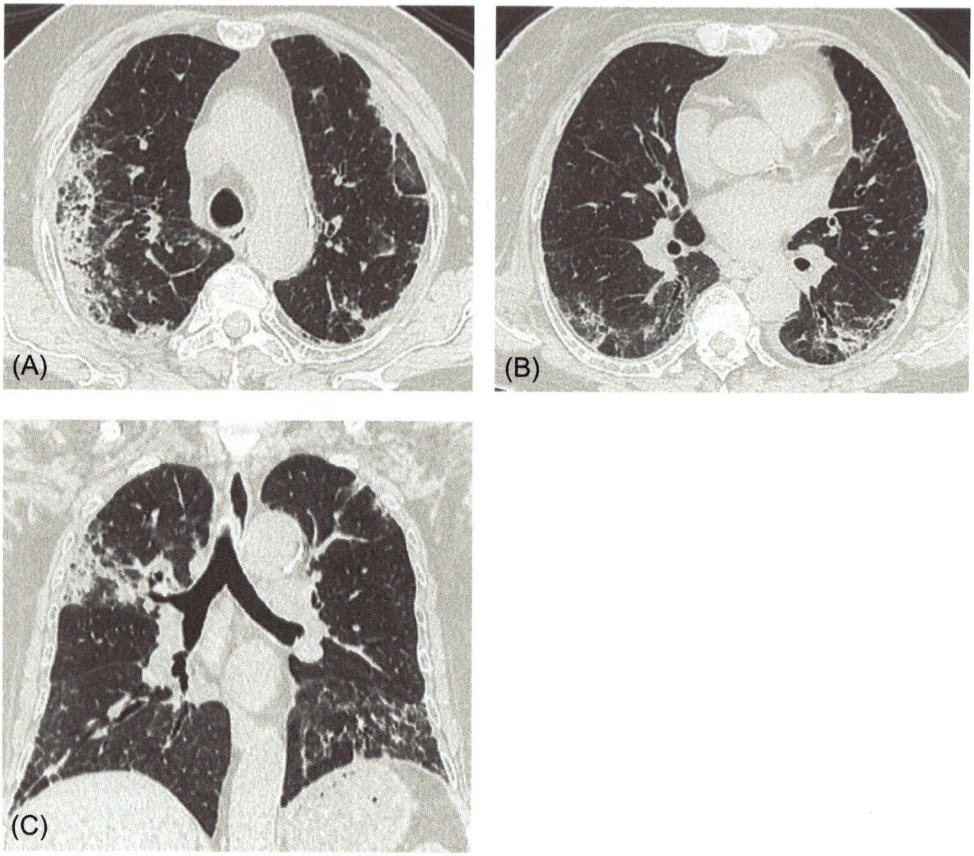

Figure 3.21 Chest computed tomography images 4 days after admission. (A) The lesions in both lungs show significant absorption, with a reduction in the area of the lesions. The bronchiectasis is more obvious. (B) The lesions in the lower lobes of both lungs are partially absorbed. (C) The coronal image shows that the lesions in the upper lobe of the right lung and the lower lobe of the left lung are partially absorbed.

Case 31

Medical history

A 56-year-old woman documented a history of contact with people who have been diagnosed with COVID-19. She reported fever 2 weeks prior to the admission, with a maximum temperature of 37.3°C, accompanied by fatigue, lethargy, and watery diarrhea. The throat swab nucleic acid test was positive 5 days prior to the date of admission. She reported chest tightness and mild dyspnea 1 day prior to the date of admission.

Laboratory findings

White blood cell count: 8.60×10^9/L, lymphocyte count: 1.09×10^9/L, lymphocyte percentage: 12.7%, C-reactive protein: 5.3 mg/L.

Radiological findings

The radiological findings of chest CT scans during her hospitalization are as follows: Figs. 3.22 and 3.23.

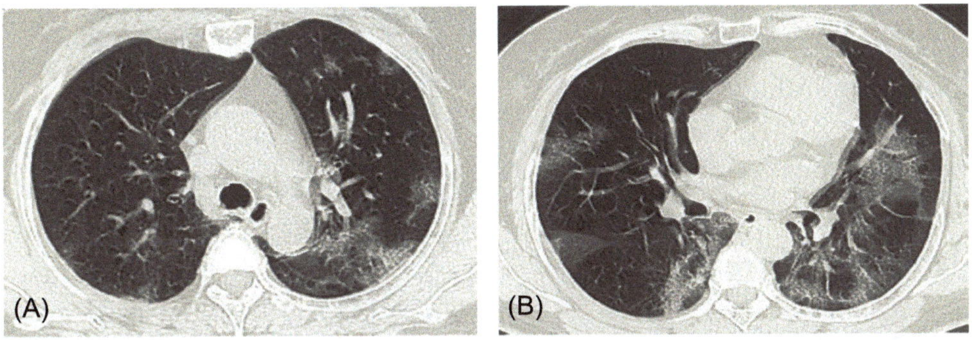

Figure 3.22 Chest computed tomography images 2 days after admission. (A) Ground-glass opacities are noted in the upper lobes of both lungs. (B) Ground-glass opacities are noted in both lungs.

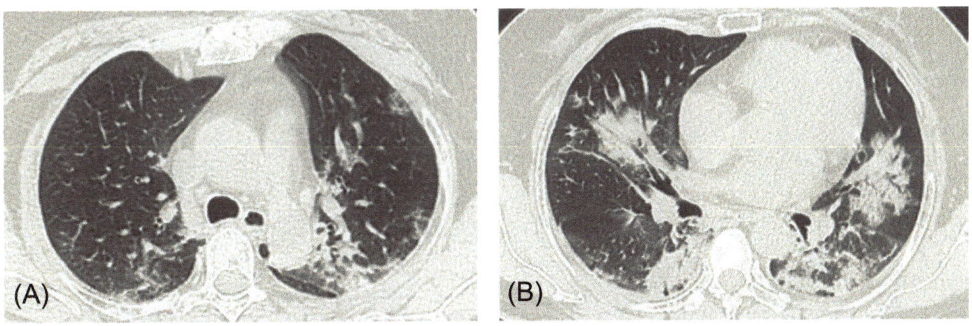

Figure 3.23 Chest computed tomography images 5 days after admission. (A) The lesions are significantly enlarged in both lungs, with most ground-glass opacities turning into consolidations and air bronchogram seen in them. (B) The lesions are significantly enlarged in both lungs, with most ground-glass opacities turning into consolidations, with air bronchogram seen in them.

Case 32

Medical history

A 65-year-old woman documented a history of contact with people who came from the epidemic area. She reported chest tightness and paroxysmal irritant dry coughing bouts, cough, fatigue, and aching limbs about 10 days prior to the date of admission. The throat swab nucleic acid test was positive 8 days prior to the date of admission.

She responded poorly to antiviral treatment. After giving antibiotics and high flow oxygen inhalation, her chest tightness and cough were relieved mildly. The throat swab nucleic acid test was negative 2 days prior to the date of admission, whereas there was no significant improvement in oxygen partial pressure and oxygenation index.

Laboratory findings

White blood cell count: 8.10×10^9/L, lymphocyte count: 2.30×10^9/L, lymphocyte percentage: 28.4%, C-reactive protein: 4.1 mg/L.

Radiological findings

The radiological findings of chest CT scans during her hospitalization are as follows: Figs. 3.24 and 3.25.

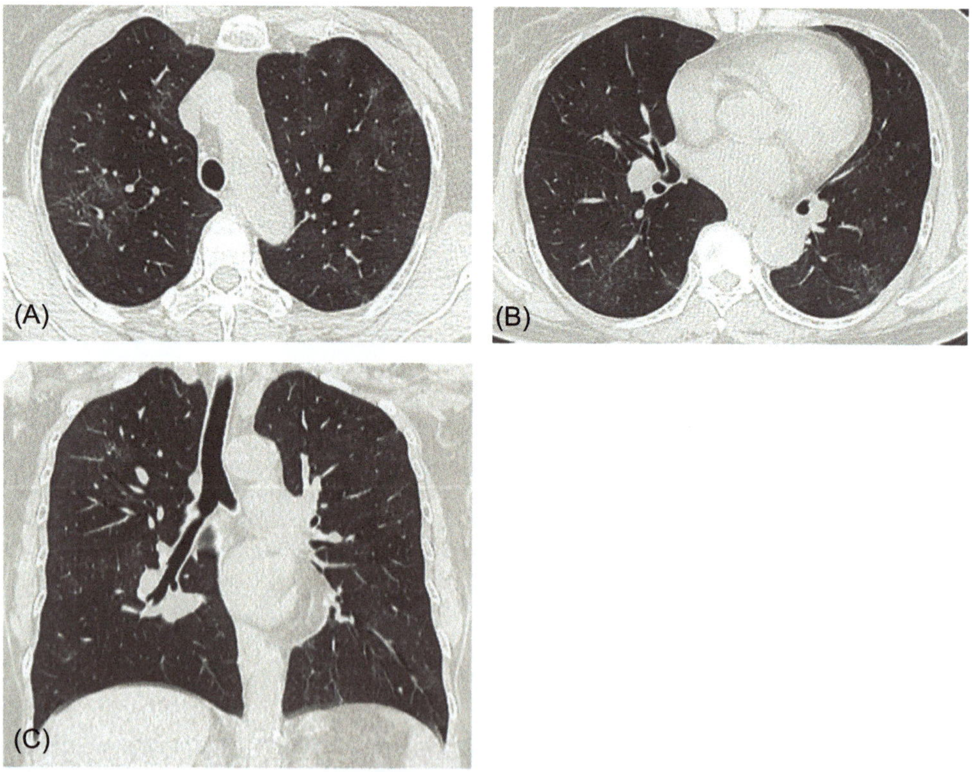

Figure 3.24 Chest computed tomography images 2 days after admission. (A) Multiple ground-glass opacities are noted in the upper lobes of both lungs. (B) Multiple diffuse ground-glass opacities in bilateral lungs are noted. (C) The coronal image shows multiple ground-glass opacities in both lungs.

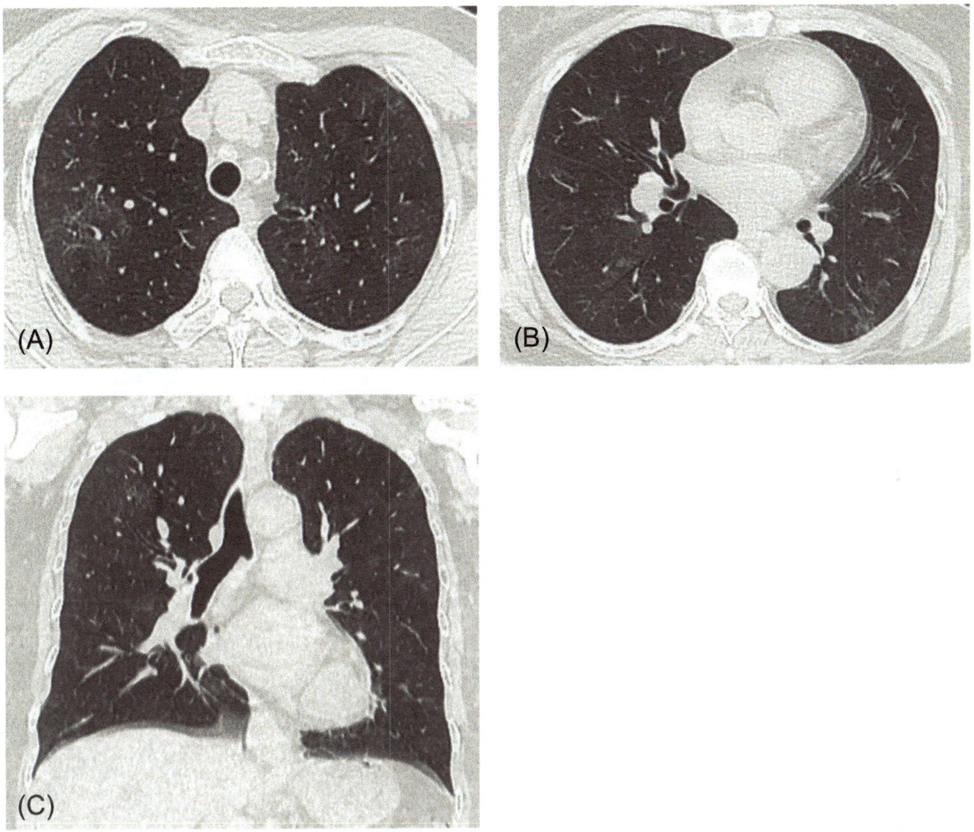

Figure 3.25 Chest computed tomography images 4 days after admission. (A) Mild reduction in the area of ground-glass opacities in bilateral lungs is noted. (B) Mild reduction in the area of ground-glass opacities in bilateral lungs is noted. (C) The coronal image shows mild reduction in the area of ground-glass opacities in bilateral lungs.

Case 33

Medical history

A 66-year-old woman documented a history of contact with people suspected of having COVID-19. She reported fever 26 days prior to the admission, with a maximum temperature of 38.0°C, accompanied by cough and expectoration with yellow sticky sputum.

Laboratory findings

White blood cell count: 7.80×10^9/L, lymphocyte count: 2.02×10^9/L, lymphocyte percentage: 25.9%, C-reactive protein: 116.5 mg/L.

Radiological findings

The radiological findings of chest CT scans during her hospitalization are as follows: Figs. 3.26 and 3.27.

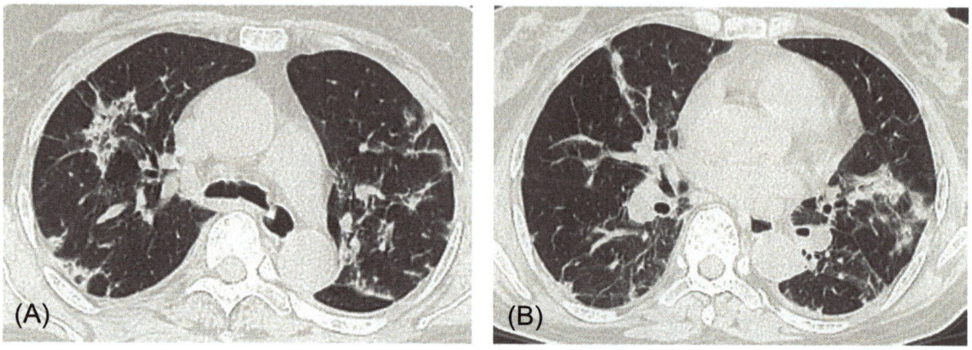

Figure 3.26 Chest computed tomography images 2 days after admission. (A) Multiple consolidations and fibrotic bands are noted in the upper lobes of both lungs. (B) Multiple lung consolidations, ground glass opacities and fibrotic bands in the upper and lower lobes of both lungs are noted.

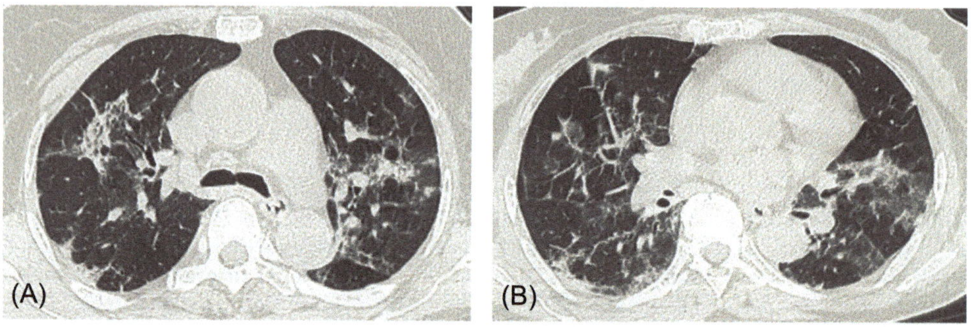

Figure 3.27 Chest computed tomography images 7 days after admission. (A) The extent of the lesion is slightly enlarged in the upper lobe of the left lung. (B) The lesions in the upper lobe of the left lung are partially absorbed, while there are new emerging ground-glass opacities in bilateral lungs.

Case 34

Medical history

A 63-year-old woman documented a history of contact with people who had had traveled recently. She suffered from irritating dry coughing bouts and was short of breath 1 month prior to the date of admission. She reported fever 21 days prior to the admission, with a maximum temperature of 37.8°C. The throat swab nucleic acid test was positive 11 days prior to the admission. Arterial blood gas indicated that the oxygenation index was less than 300 just three days prior to the admission. The patient reported chest tightness, shortness of breath, cough, and expectoration.

Laboratory findings

White blood cell count: 5.50×10^9/L, lymphocyte count: 1.41×10^9/L, lymphocyte percentage: 25.7%, C-reactive protein: 12.7 mg/L.

Radiological findings

The radiological findings of chest CT scans during her hospitalization are as follows: Fig. 3.28.

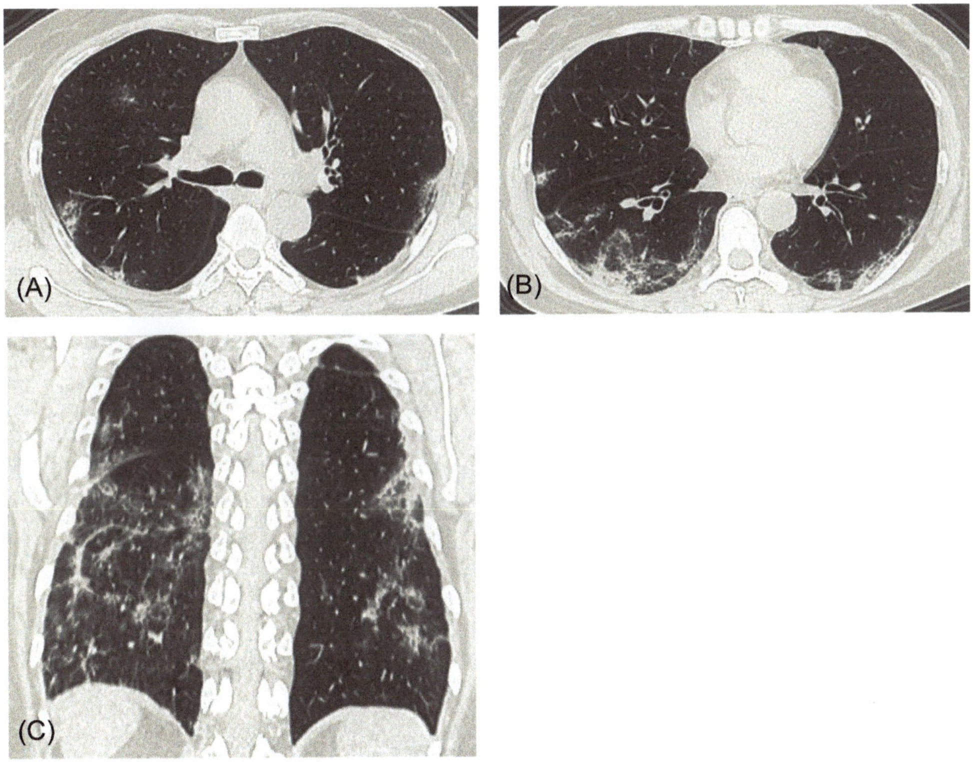

Figure 3.28 Chest computed tomography images 2 days after admission. (A) Multiple ground glass opacities in the upper lobes of both lungs are noted. (B) There are multiple ground-glass opacities, consolidations and bands in bilateral lungs, predominantly in the subpleural area. (C) The coronal image show multiple consolidations and ground-glass opacities and fibrotic bands in both lungs.

Case 35

Medical history

The patient is a 66-year-old man. He documented a history of contact with people who were diagnosed with COVID-19. He started to have coughing bouts with yellow sticky sputum 2 weeks prior to the date of admission. The throat swab nucleic acid test was positive 13 days prior to the date of admission. The nucleic acid test was still positive after receiving antiviral treatment.

Laboratory findings

White blood cell count: $3.20 \times 10^9/L$, lymphocyte count: $0.90 \times 10^9/L$, lymphocyte percentage: 28.2%, C-reactive protein: 21.9 mg/L.

Radiological findings

The radiological findings of chest CT scans during his hospitalization are as follows: Figs. 3.29–3.31.

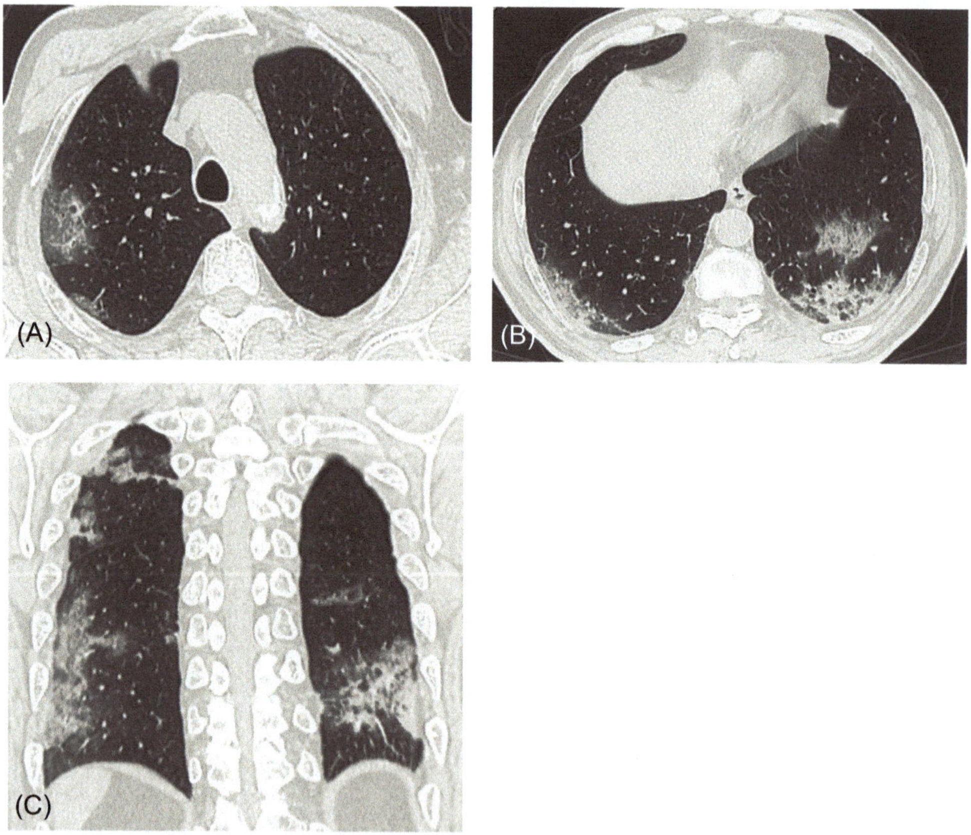

Figure 3.29 Chest computed tomography images 2 days after admission. (A) Ground-glass opacities in the upper lobe of the right lung are noted. (B) Multiple consolidations and ground-glass opacities in bilateral lower lobes are noted. (C) The coronal image shows multiple consolidations and ground glass opacities in both lungs.

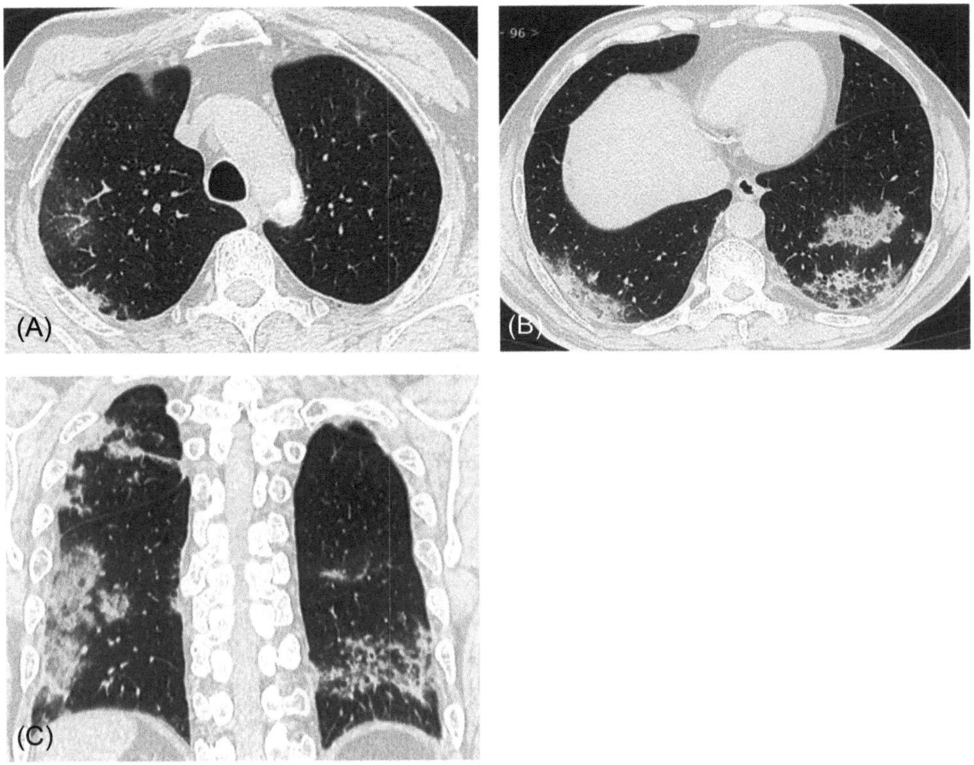

Figure 3.30 Chest computed tomography images 5 days after admission. (A) Ground-glass opacities and consolidations are noted in the upper lobe of the right lung. A few ground-glass opacities are noted in the upper lobe of the left lung. (B) The lesions in the right lung's lower lobe show sustained attenuation. (C) The coronal image show multiple consolidations in both lungs, which are more dense than before.

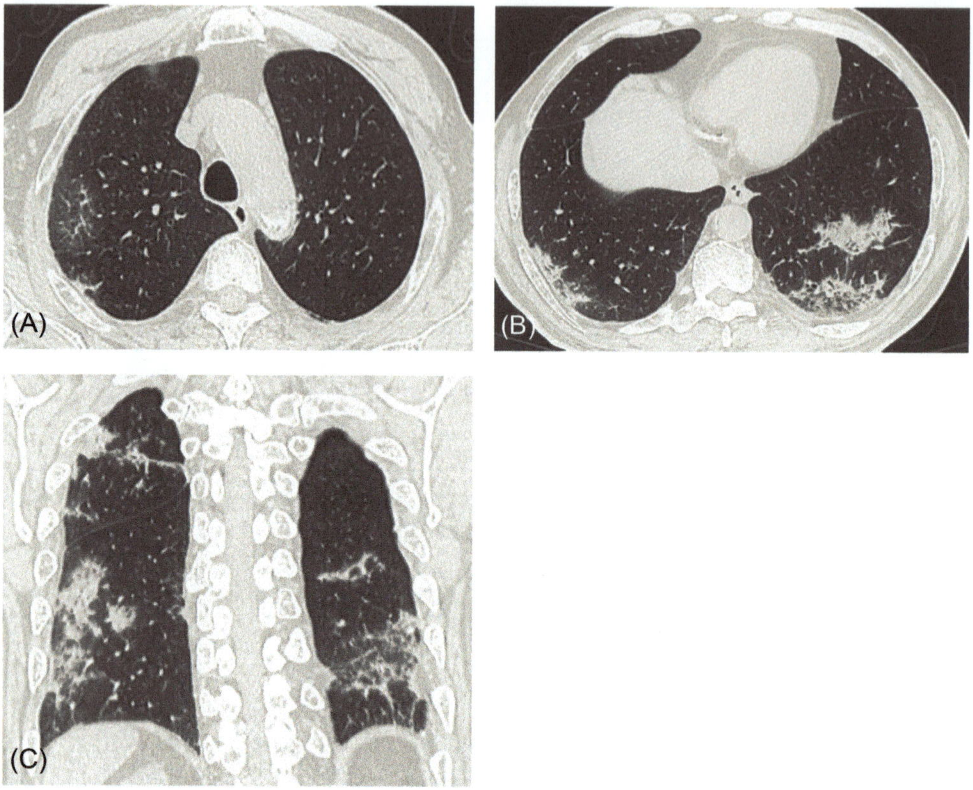

Figure 3.31 Chest computed tomography images 8 days after admission. (A) The lesions in the upper lobe of the right lung are partially absorbed, with reduced extent and attenuation. The upper lobe lesion of the left lung is almost completely absorbed. (B) The lesions in the lower lobes of both lungs are partially absorbed. (C) The coronal image shows partial absorption of the lesions in the two lungs.

Case 36

Medical history

The patient is a 31-year-old woman. She worked in the infection ward just 13 days prior to the date of admission. She reported fever 2 days prior to the date of admission with a maximum temperature of 37.5°C, accompanied by coughing bouts, dry cough, and mild fatigue. The nucleic acid test was positive 1 day prior to the date of admission.

Laboratory findings

White blood cell count: 3.20×10^9/L, lymphocyte count: 1.88×10^9/L, lymphocyte percentage: 58.9%, C-reactive protein: 35.5 mg/L.

Radiological findings

The radiological findings of the chest CT scans during her hospitalization are as follows: Figs. 3.32 and 3.33.

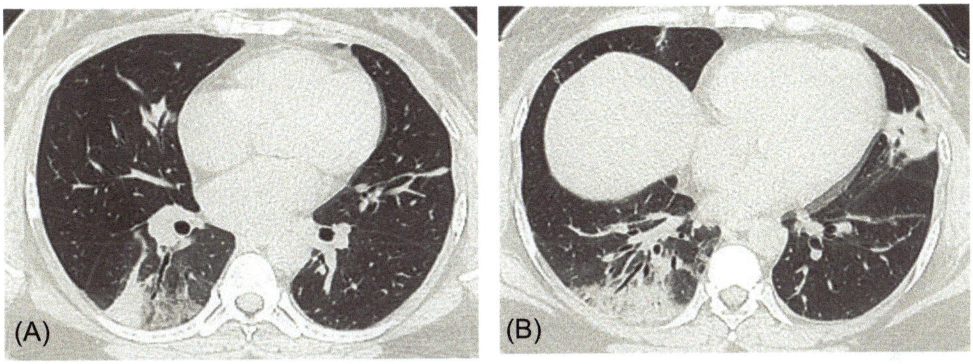

Figure 3.32 Chest computed tomography images 2 days after admission. (A) Consolidation is noted in the upper lobes of both lungs, consolidations and ground-glass opacities are noted in the lower lobe of the right lung. (B) There is consolidation in the lingual segment of the left lung's upper lobe. There are ground-glass opacities and consolidations in the right lung's lower lobe.

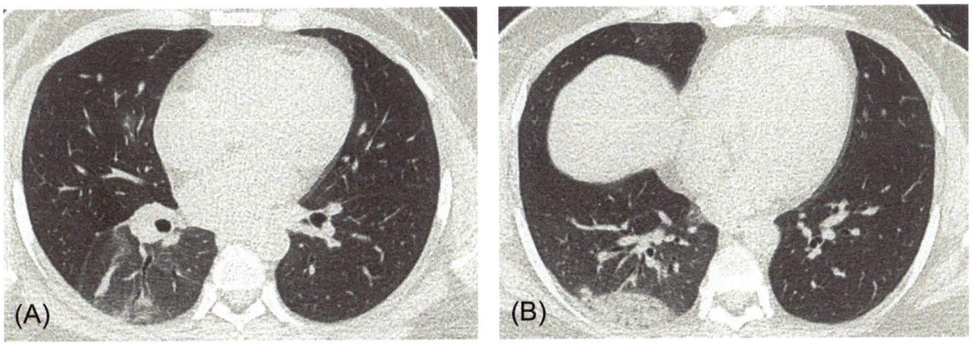

Figure 3.33 Chest computed tomography images 5 days after admission. (A) The consolidation in the lingual segment of left lung's upper lobe is completely absorbed. There is a significant reduction in the area and attenuation of the lesions in the right lung. (B) The consolidation in the lingual segment of left lung's upper lobe is completely absorbed. There is a significant reduction in the area and attenuation of the lesions in the right lung.

Case 37

Medical history

A 57-year-old man had diarrhea 18 days prior to the admission, with 3 to 4 times of watery stool per day. The symptoms worsened 13 days prior to the date of admission. The throat swab nucleic acid test was positive 12 days prior to the admission.

Laboratory findings

White blood cell count: 7.40×10^9/L, lymphocyte count: 2.04×10^9/L, lymphocyte percentage: 27.4%, C-reactive protein: 7.8 mg/L.

Radiological findings

The radiological findings of chest CT scans during his hospitalization are as follows: Figs. 3.34 and 3.35.

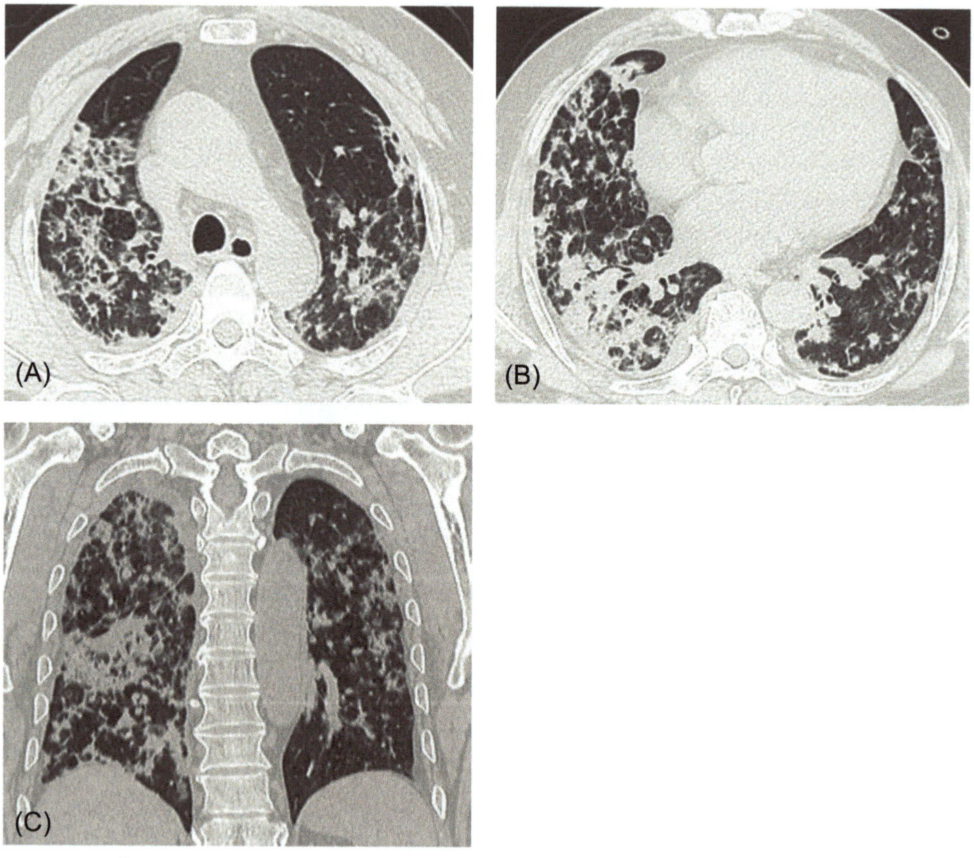

Figure 3.34 Chest computed tomography images 2 days after admission. (A) Multiple consolidations, ground glass opacities, fibrotic bands and some bronchiectasis in the upper lobes of both lungs are noted. (B) Multiple consolidations, ground glass opacities, fibrotic bands and some bronchiectasis in both lungs are noted. (C) Multiple lung consolidations, ground-glass opacities, fibrotic bands and some bronchiectasis in both lungs are noted.

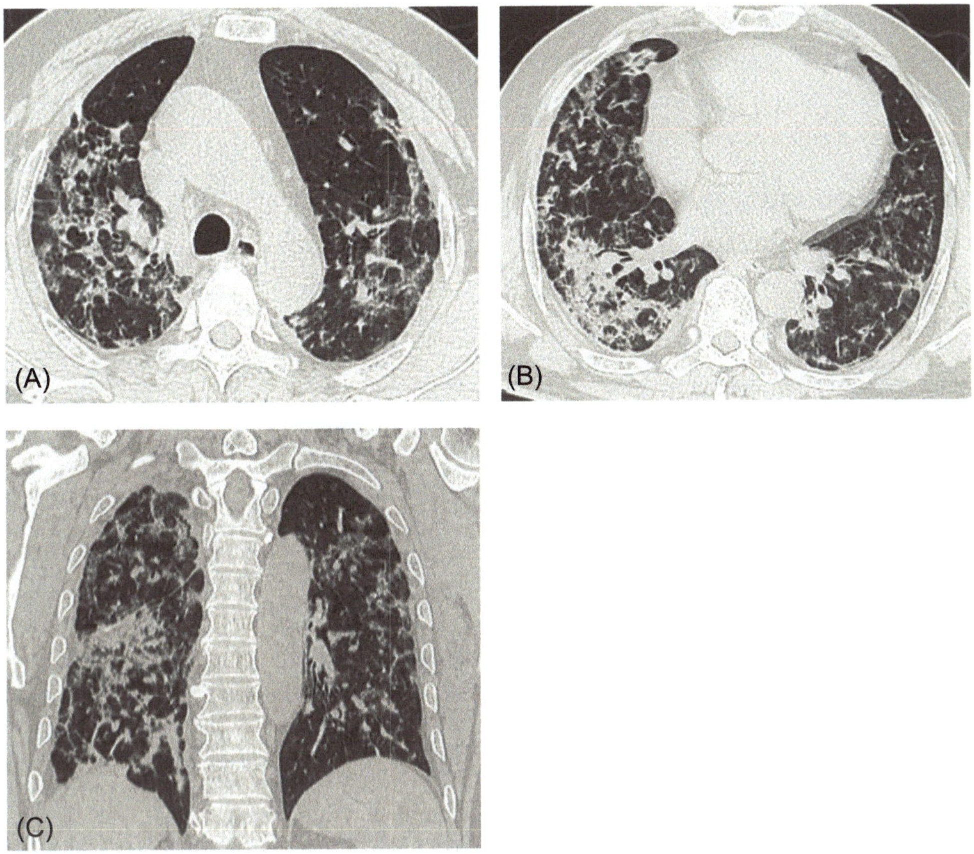

Figure 3.35 Chest computed tomography images 7 days after admission. (A) The lesions in the upper lobes of both lungs are partially absorbed. (B) The lesions in both lungs are partially absorbed. (C) The lesions in both lungs are partially absorbed.

Case 38

Medical history

The patient is a 70-year-old woman. She reported fever with a maximum temperature of 38.1°C just 34 days prior to the date of admission, accompanied by cough, expectoration with white sticky sputum, chills, fatigue, poor appetite, headache, and nausea. She had close contact with her brother (who came from the epidemic area and was diagnosed with COVID-19) 24 days prior to the date of admission. The throat swab nucleic acid test was positive 23 days prior to the date of admission. She received antiviral and traditional Chinese medicine therapy before she again underwent the throat swab nucleic acid test, which was still positive 11 days prior to the date of admission. The symptoms of cough and expectoration worsened, accompanied by chest tightness and dyspnea.

Laboratory findings

White blood cell count: 4.30×10^9/L, lymphocyte count: 1.04×10^9/L, lymphocyte percentage: 24.1%, C-reactive protein: 12.4 mg/L.

Radiological findings

The radiological findings of chest CT scans during her hospitalization are as follows: Figs. 3.36–3.38.

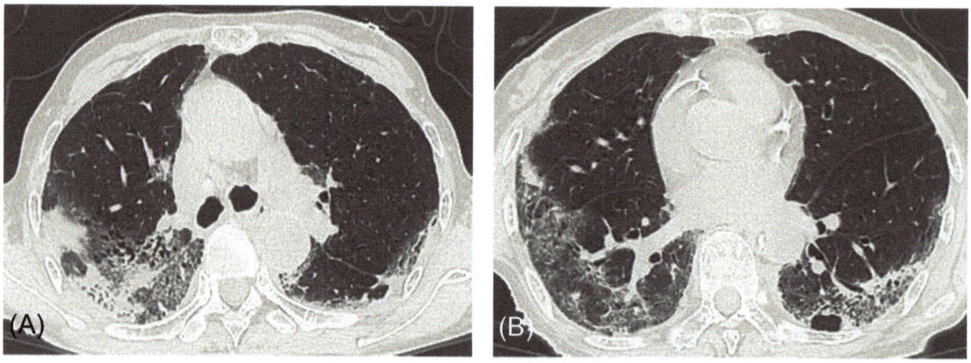

Figure 3.36 Chest computed tomography images 8 days after admission. (A) There are multiple ground-glass opacities in the bilateral lungs, predominant in the bilateral lower lobes, with thickening of intra- and interlobular septa, consolidations and bronchiectasis seen in the lesions. (B) There is a bulla with thick wall in the subpleural area of the left lung's lower lobe.

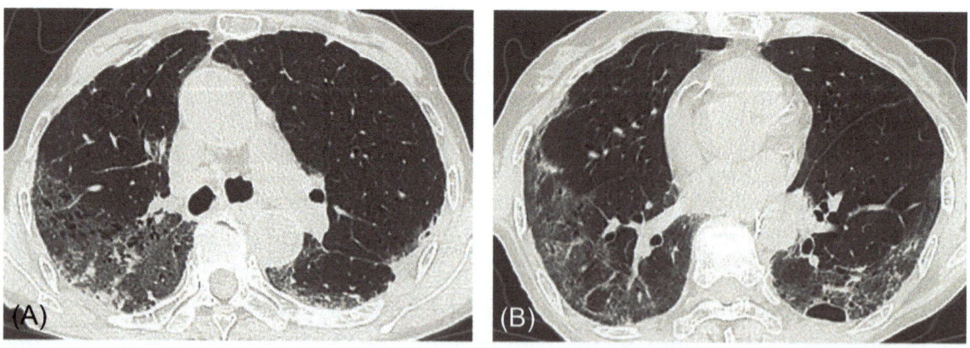

Figure 3.37 Chest computed tomography images 12 days after admission. (A) The lesions in the upper and lower lobes of both lungs are partially absorbed, with reduced extent and attenuation. (B) The lesions in the lower lobes of both lungs and the upper lobe of the right lung are partially absorbed, with reduced extent and attenuation.

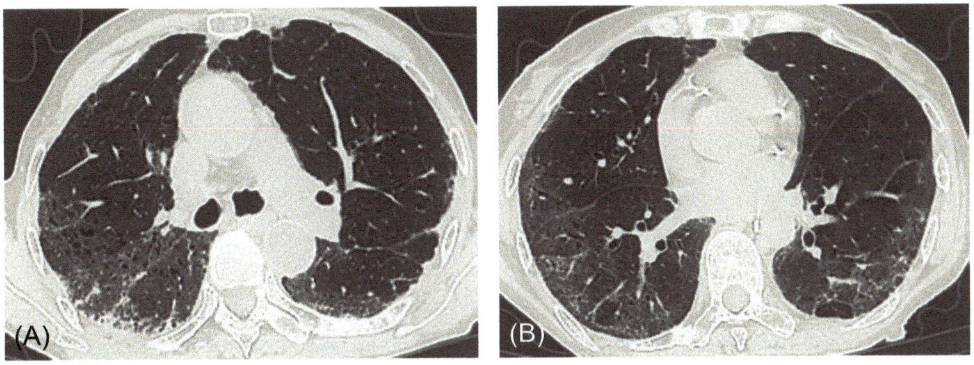

Figure 3.38 Chest computed tomography images 20 days after admission. (A) The absorption continues for the lesions in both the lungs, with a remarkable reduction in the area of the lesions. (B) The absorption continues for the lesions in both the lungs, with a remarkable reduction in the area of the lesions.

Case 39

Medical history

The patient is a 26-year-old woman. She documented a history of contact with people who had had been diagnosed with COVID-19. She had nausea and vomiting 11 days prior to the date of admission and was treated as gastritis by the local clinic with poor response. She reported fever 7 days prior to the date of admission, with a maximum temperature of 37.5°C, accompanied by cough, expectoration with white sticky sputum, weakness, and fatigue.

Laboratory findings

White blood cell count: 10.40×10^9/L, lymphocyte count: 2.07×10^9/L, lymphocyte percentage: 19.9%, C-reactive protein: 2.2 mg/L.

Radiological findings

The radiological findings of chest CT scans during her hospitalization are as follows: Figs. 3.39–3.41.

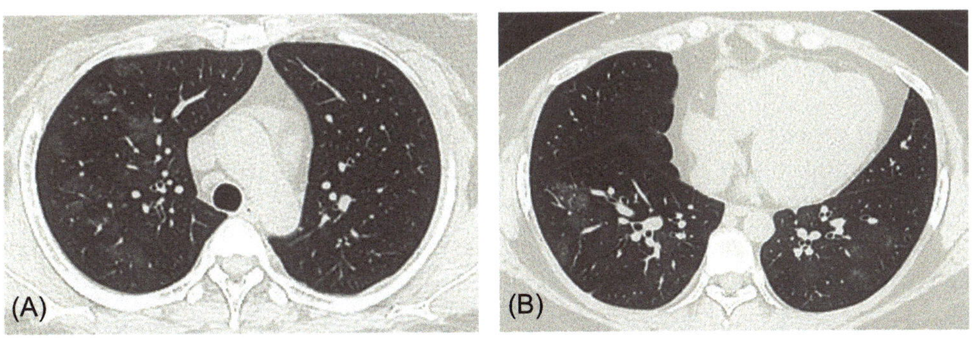

Figure 3.39 Chest computed tomography images 2 days after admission. (A) Multiple ground glass opacities in the upper lobe of the right lung are noted. (B) Multiple ground glass opacities in the lower lobes of both lungs are noted.

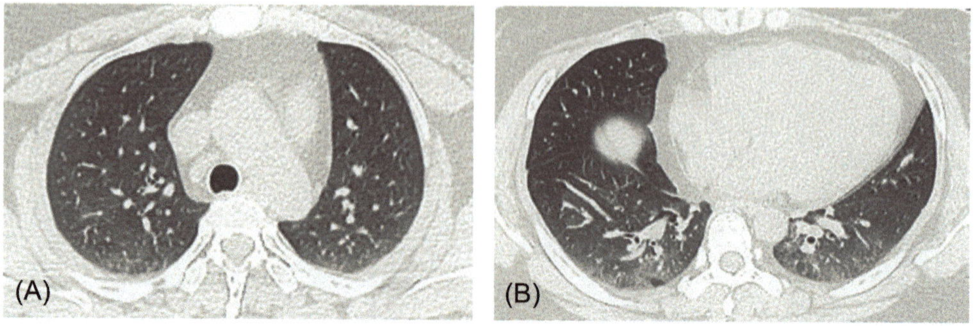

Figure 3.40 Chest computed tomography images 8 days after admission. (A) There is obvious absorption of the lesions in the right lung's upper lobe. (B) The ground-glass opacities in the bilateral lower lobes are enlarged.

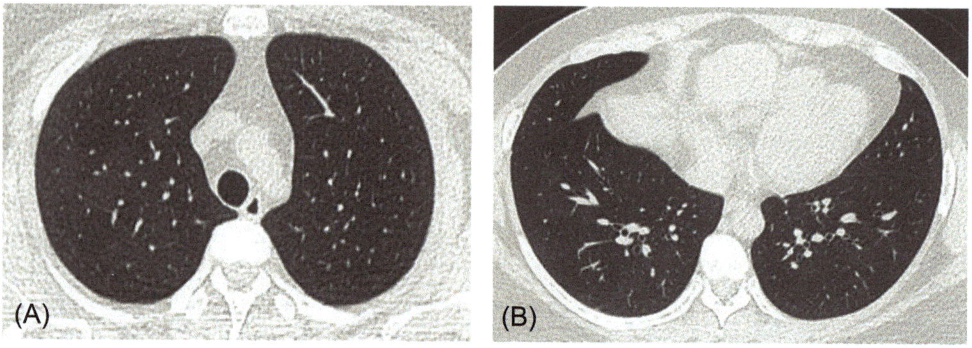

Figure 3.41 Chest computed tomography images 11 days after admission. (A) The upper lobe lesions of the right lung are completely absorbed. (B) The lesions in the lower lobes of both lungs are completely absorbed.

Case 40

Medical history

A 46-year-old woman took care of patients during the outbreak of COVID-19. She had dry throat 11 days prior to the date of admission and reported fever 8 days prior to the admission date, with a maximum temperature of 38.5°C. The throat swab nucleic acid test was positive 6 days prior to the admission date. The symptoms worsened 2 days prior to the date of admission, with a sustained fever of 38.2°C, accompanied by dry cough, chest tightness, and asthma.

Laboratory findings

White blood cell count: 4.80×10^9/L, lymphocyte count: 0.94×10^9/L, lymphocyte percentage: 19.5%, C-reactive protein: 72.4 mg/L.

Radiological findings

The radiological findings of chest CT scans during her hospitalization are as follows: Figs. 3.42–3.44.

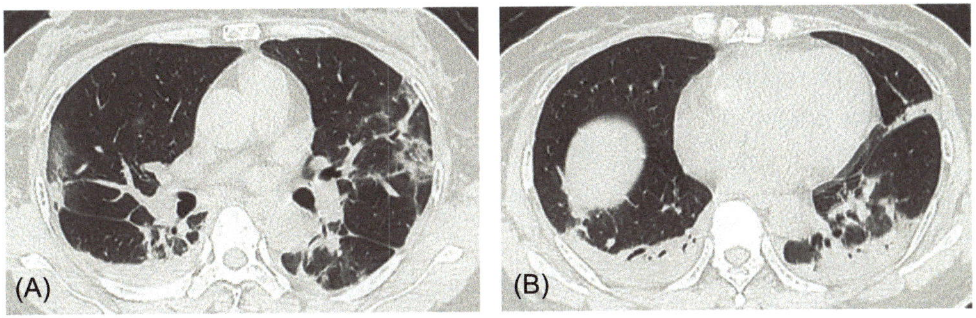

Figure 3.42 Chest computed tomography images 2 days after admission. (A) There are multiple ground-glass opacities, consolidations and bands in the bilateral upper and lower lobes. (B) There are multiple ground-glass opacities, consolidations and bands in the bilateral lungs, with air bronchogram seen in the subpleural consolidative lesions.

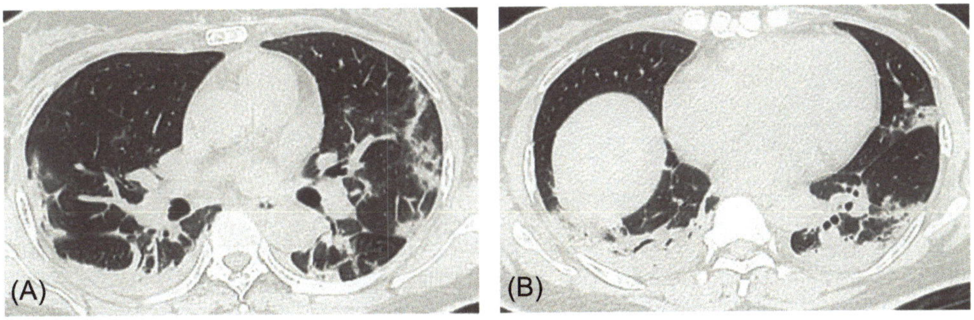

Figure 3.43 Chest computed tomography images 5 days after admission. (A) Some of the ground-glass opacities have turned into consolidations, with the other lesions remaining similar. (B) The lesion in the lingual segment of the left upper lobe enlarges in size.

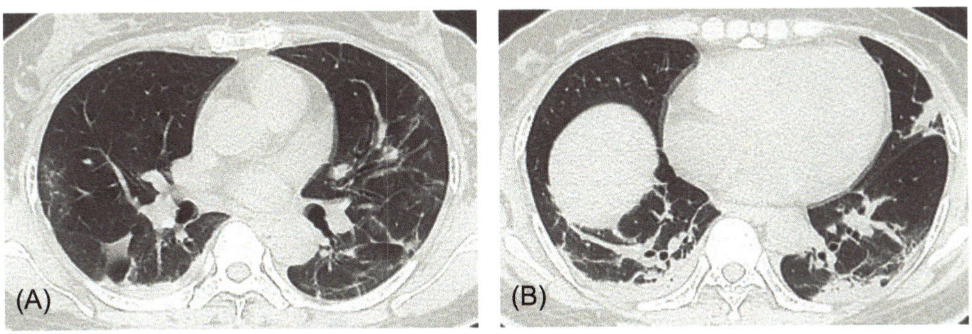

Figure 3.44 Chest computed tomography images 8 days after admission. (A) The lesions in the upper and lower lobes of both lungs are partially absorbed. (B) The lesions in the lower lobe of both lungs are partially absorbed.

Case 41

Medical history

The patient is a 59-year-old man. He documented a history of contact with people who came from the epidemic area. He reported fever 20 days prior to the date of admission, with a maximum temperature of 37.5°C. The throat swab nucleic acid test was positive 16 days prior to the admission date.

Laboratory findings

White blood cell count: $5.70 \times 10^9/L$, lymphocyte count: $0.88 \times 10^9/L$, lymphocyte percentage: 15.3%, C-reactive protein: 40.0 mg/L.

Radiological findings

The radiological findings of chest CT scans during his hospitalization are as follows: Figs. 3.45–3.47.

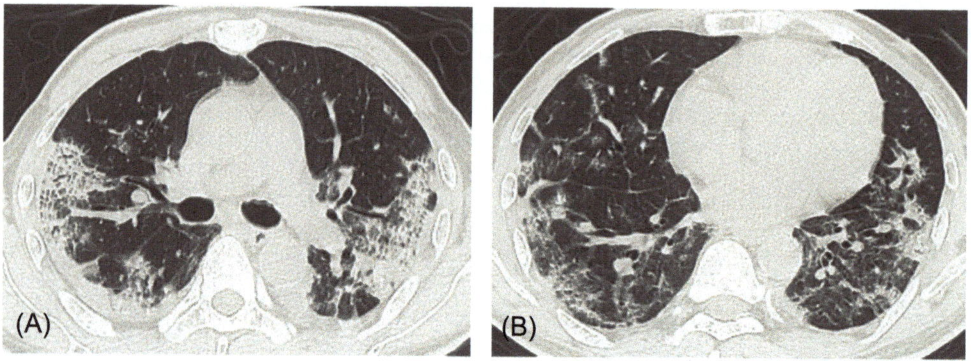

Figure 3.45 Chest computed tomography images 2 days after admission. (A) There are multiple consolidations and ground glass opacities in the upper and lower lobes of both lungs. (B) There are multiple consolidations, ground glass opacities and fibrotic bands in both lungs.

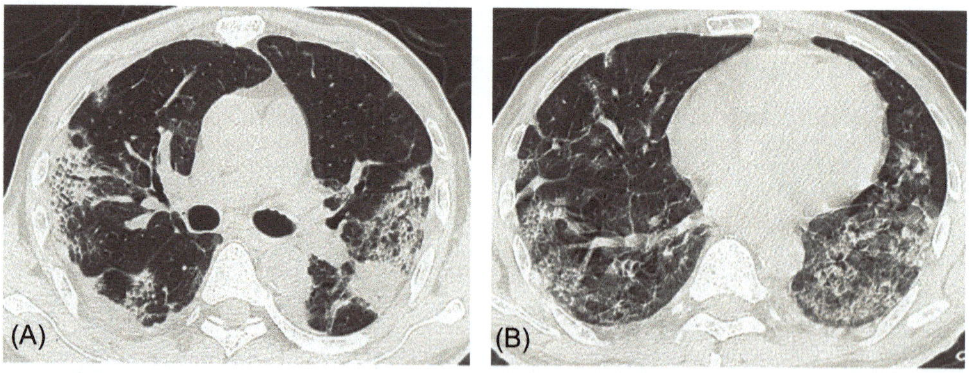

Figure 3.46 Chest computed tomography images 4 days after admission. (A) There is a mild reduction in the area of the lesions in the right lung. (B) Consolidations enlarge in the area in the left lung, and the ground-glass opacities and bands remain similar to those of the prior images.

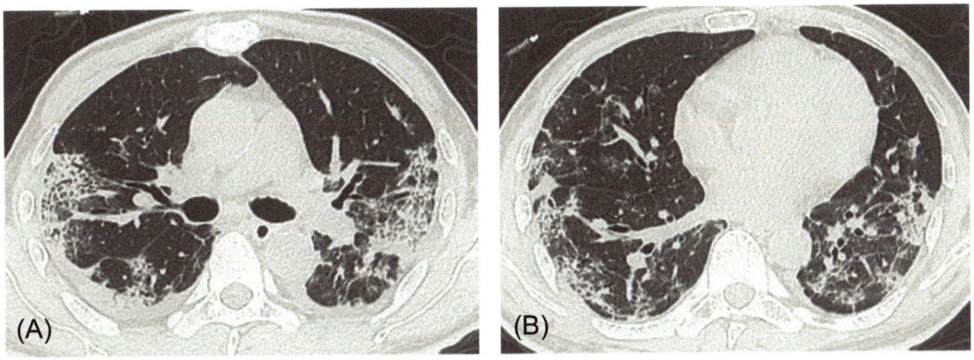

Figure 3.47 Chest computed tomography images 7 days after admission. (A) The lesions in the upper and lower lobes of both lungs are partially absorbed. (B) The lesion in the upper and lower lobes of both lungs are partially absorbed.

Case 42

Medical history

A 29-year-old woman worked in a hospital during the outbreak of COVID-19. She reported fever 4 days prior to the date of admission, with a maximum temperature of 37.5°C. The throat swab nucleic acid test was positive 3 days prior to the admission date. After receiving antiviral treatment, her temperature returned to normal while she suffered from diarrhea and weakness.

Laboratory findings

White blood cell count: $5.20 \times 10^9/L$, lymphocyte count: $2.10 \times 10^9/L$, lymphocyte percentage: 40.8%, C-reactive protein: 8.5 mg/L.

Radiological findings

The radiological findings of chest CT scans during his hospitalization are as follows: Figs. 3.48–3.50.

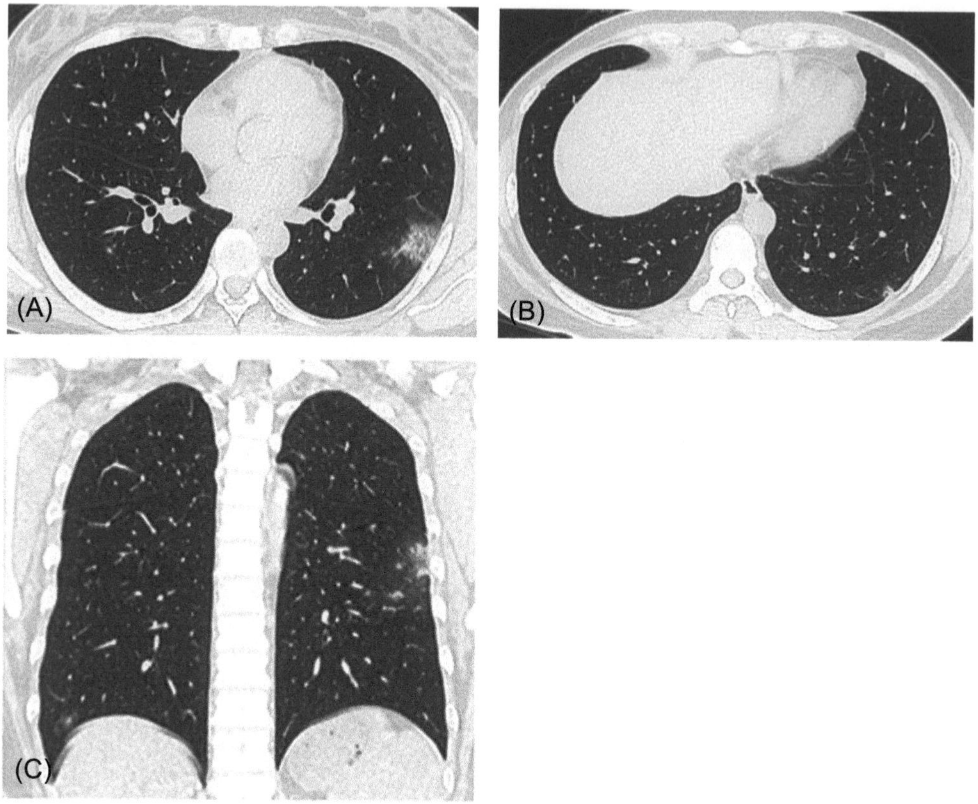

Figure 3.48 Chest computed tomography images on the day of admission. (A) There are a few ground glass opacities in the lower lobes of both lungs. (B) There is a ground-glass opacity in the lower lobe of the left lung. (C) The coronal image shows a few ground glass opacities in the lower lobes of both lungs.

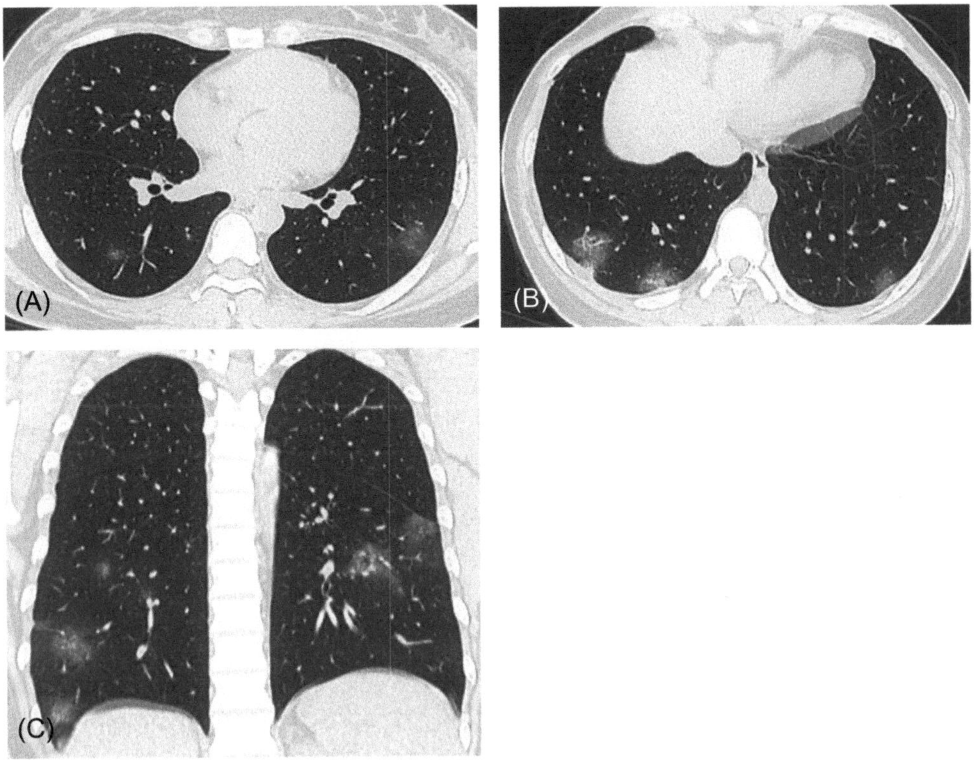

Figure 3.49 Chest computed tomography images 4 days after admission. (A) Multiple new emerging ground-glass opacities are noted in the bilateral lower lobes. (B) Multiple new emerging ground-glass opacities are noted in the bilateral lower lobes. (C) The coronal image shows multiple new emerging ground-glass opacities in the bilateral lower lobes.

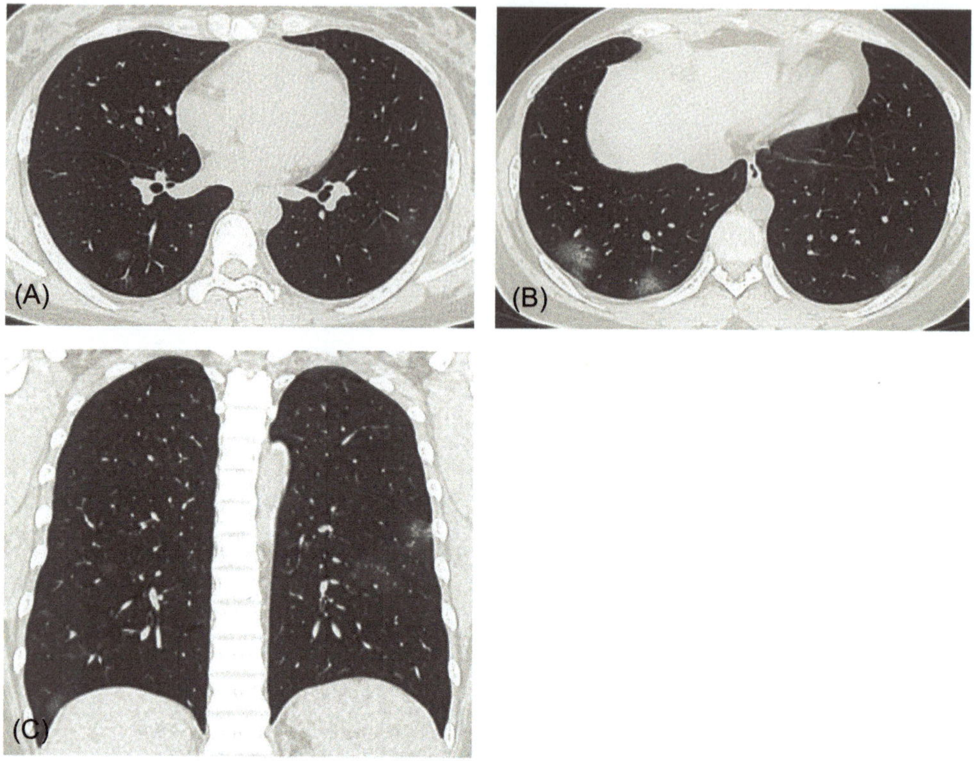

Figure 3.50 Chest computed tomography images 7 days after admission. (A) The lesions in the lower lobes of both lungs are obviously absorbed. (B) The lesions in the lower lobes of both lungs are obviously absorbed. (C) The coronal image shows obvious absorption for the lesions in the bilateral lower lobes.

Case 43

Medical history

A 56-year-old woman documented a history of contact with people diagnosed with COVID-19. She reported fever 22 days prior to the date of admission, with a maximum temperature of 38.9°C, accompanied by gradually developing a paroxysmal dry cough. The throat swab nucleic acid test was positive 8 days prior to the admission date.

Laboratory findings

White blood cell count: 3.80×10^9/L, lymphocyte count: 1.38×10^9/L, lymphocyte percentage: 36.0%, C-reactive protein: 11.0 mg/L.

Radiological findings

The radiological findings of chest CT scans during her hospitalization are as follows: Figs. 3.51 and 3.52.

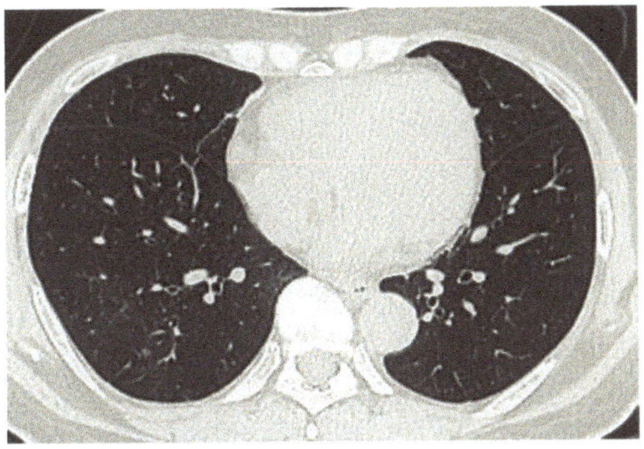

Figure 3.51 Chest computed tomography image 2 days after admission. There is a small ground-glass opacity in the right lung's lower lobe.

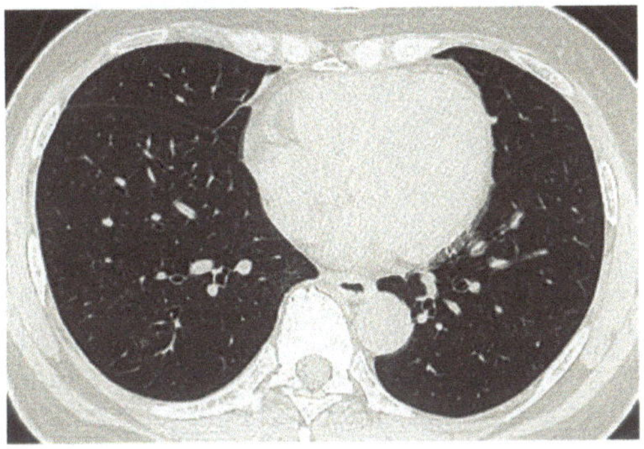

Figure 3.52 Chest computed tomography image 6 days after admission date. Compared to the chest computed tomography image of 2 days after admission, it is similar in the size of the small ground-glass opacity in the right lung's lower lobe, while the attenuation is reduced significantly.

Section 2: Manifestations of computed tomography, PET/computed tomography, and PET/MR and prognosis of severe COVID-19 cases

Case 44

Medical history

The patient is an 88-year-old woman. She documented a history of contact with people diagnosed with COVID-19. She reported fever with a maximum temperature of 38.0°C just 2 weeks prior to the admission date, accompanied by pharyngeal pain, a runny nose, chills, and general weakness. The throat swab nucleic acid test was positive 10 days prior to the date of admission.

Laboratory findings

White blood cell count: $2.60 \times 10^9/L$, lymphocyte count: $0.82 \times 10^9/L$, lymphocyte percentage: 31.2%, C-reactive protein: 1.7 mg/L.

Radiological findings

The radiological findings of chest CT scans during her hospitalization are as follows: Figs. 3.53 and 3.54.

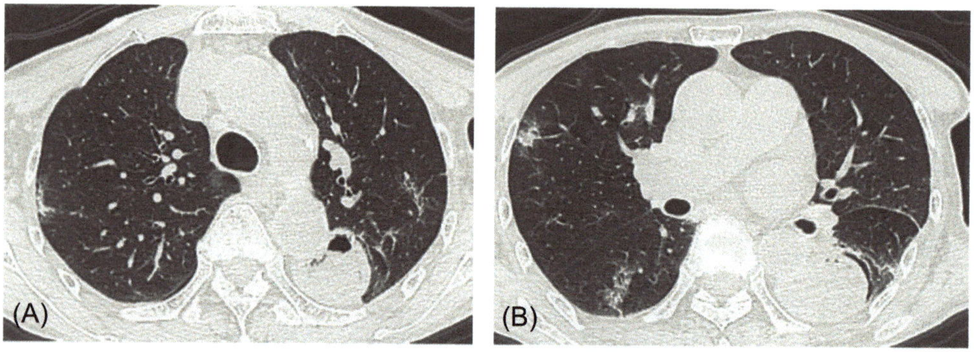

Figure 3.53 Chest computed tomography images 2 days after admission. (A) A few ground glass opacities and consolidations in the upper lobe of both lungs are noted. (B) Ground glass opacities, consolidations and fibrotic bands in the upper lobe of right lung and lower lobe of both lungs are noted.

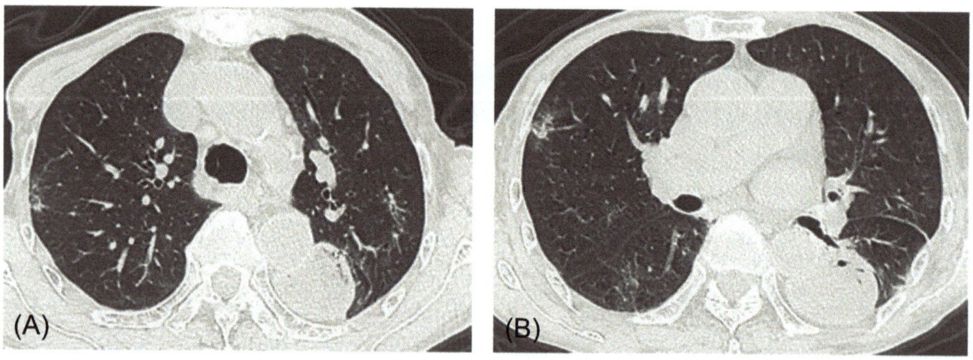

Figure 3.54 Chest computed tomography images 7 days after admission. (A) The lesions in the upper lobe of both lungs are partially absorbed. (B) The lesions in the upper lobe of the right lung and the lower lobes of both lungs are partially absorbed.

The patient underwent ^{18}F-FDG Positron Emission Tomography (PET)/CT just 8 days after the admission date, and the image manifestations are as follows: Figs. 3.55–3.62.

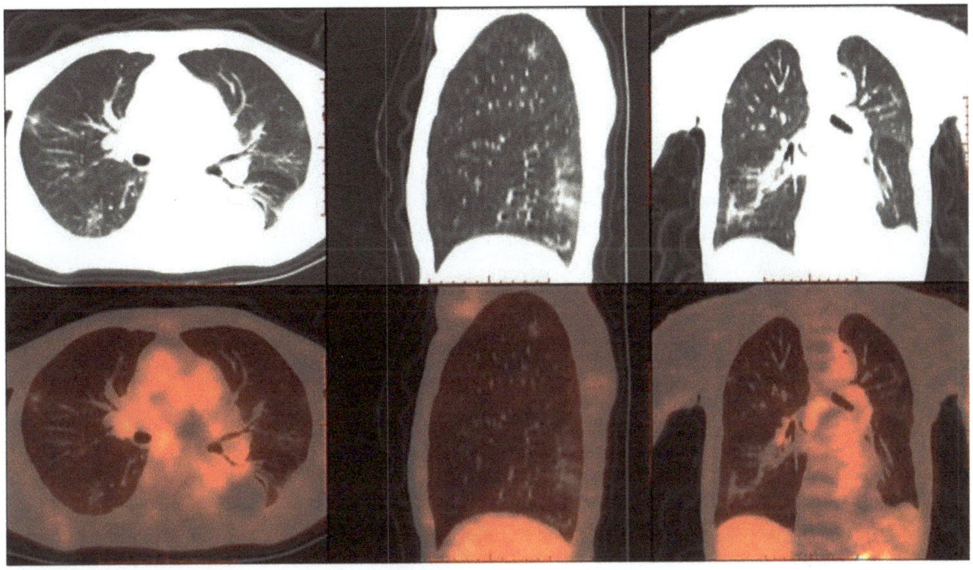

Figure 3.55 Computed tomography (top panel), PET/computed tomography fusion images (bottom panel) and 3D reconstruction images (lung window). There are multiple ground-glass and consolidative opacities, inter- and intra-lobular septal thickening, and parenchymal bands in both lungs without increased metabolic activity, suggesting a diagnosis of absorption period of COVID-19, with no metabolic activity for all lesions. *PET*, Positron Emission Tomography.

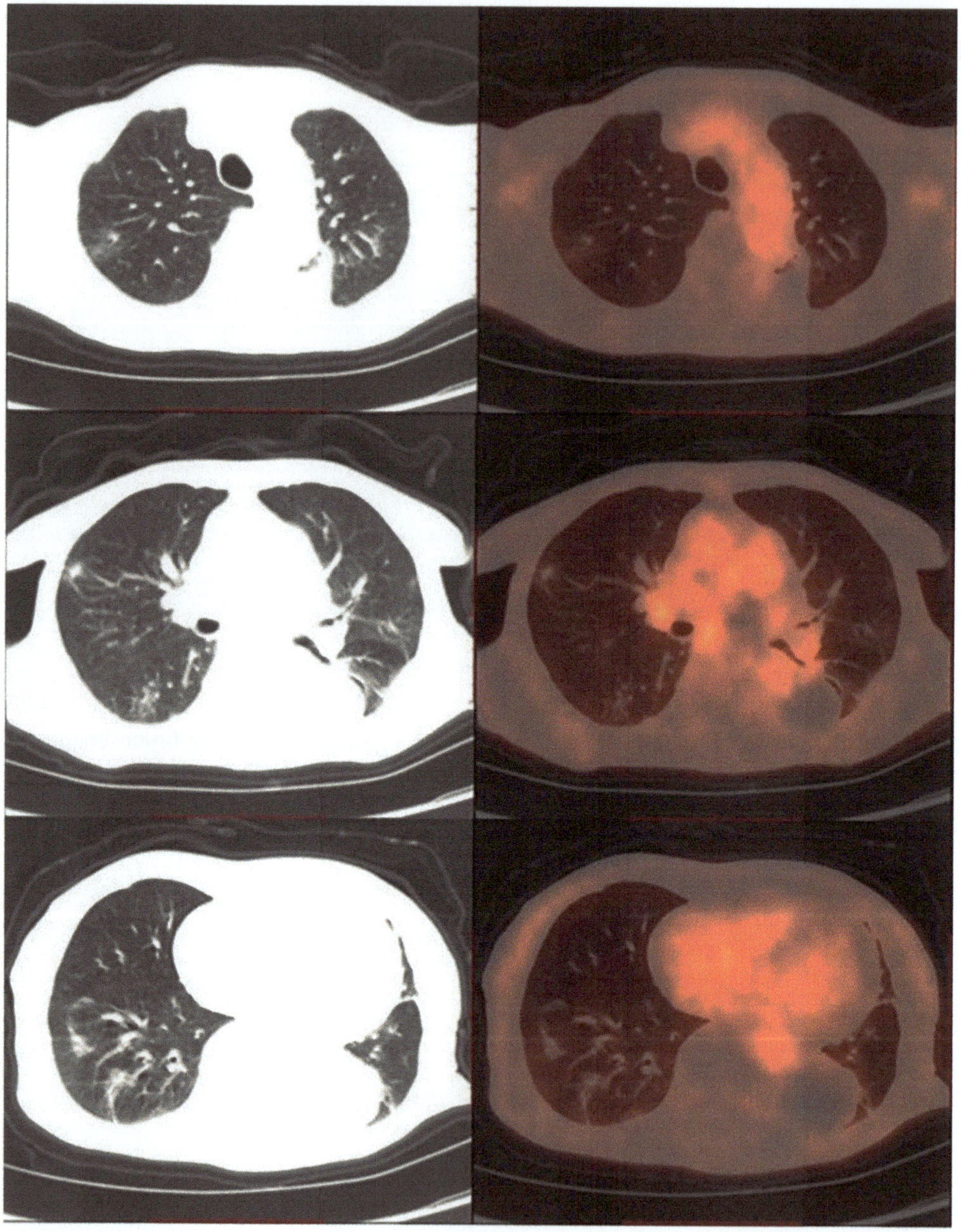

Figure 3.56 Computed tomography (left panel) and PET/computed tomography (right panel) fusion images. There are multiple ground-glass and consolidative opacities, inter- and intra-lobular septal thickening, and parenchymal bands in both lungs without increased metabolic activity, suggesting a diagnosis of absorption period of COVID-19, with no metabolic activity for all lesions. *PET*, Positron Emission Tomography.

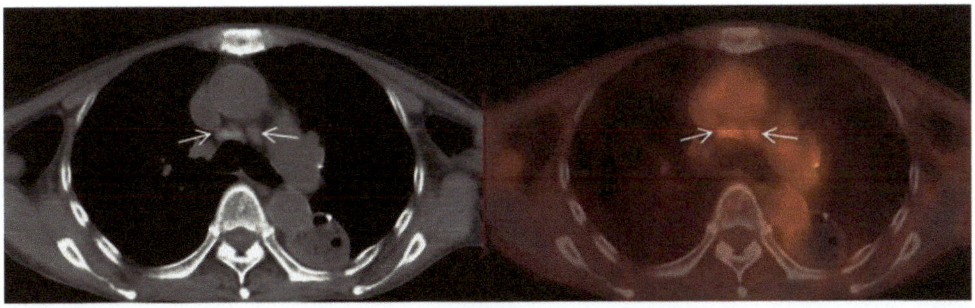

Figure 3.57 Computed tomography (left) and PET/computed tomography fusion images (right). There are multiple normal-sized lymph nodes located at the aortopulmonary level (white arrows). They show high density on CT and increased metabolic activity, suggesting a diagnosis of inflammatory lymph nodes. *PET*, Positron Emission Tomography.

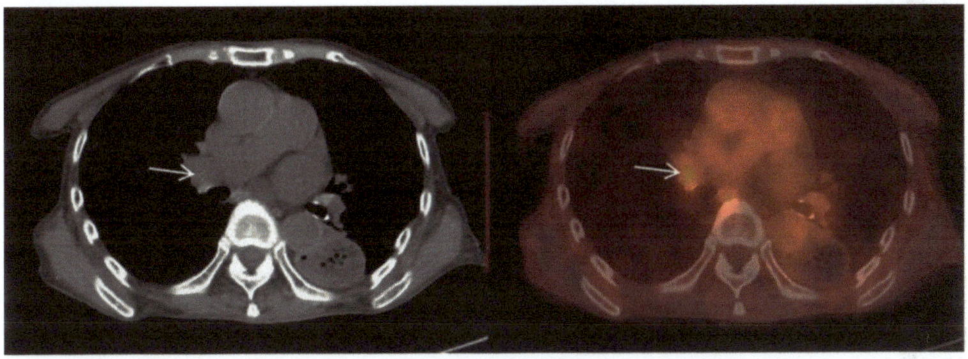

Figure 3.58 Computed tomography (left) and PET/computed tomography fusion images (right). There are multiple normal-sized lymph nodes located in both pulmonary hilum areas (white arrows). They show isodense on CT and increased metabolic activity, suggesting a diagnosis of inflammatory lymph nodes. *PET*, Positron Emission Tomography.

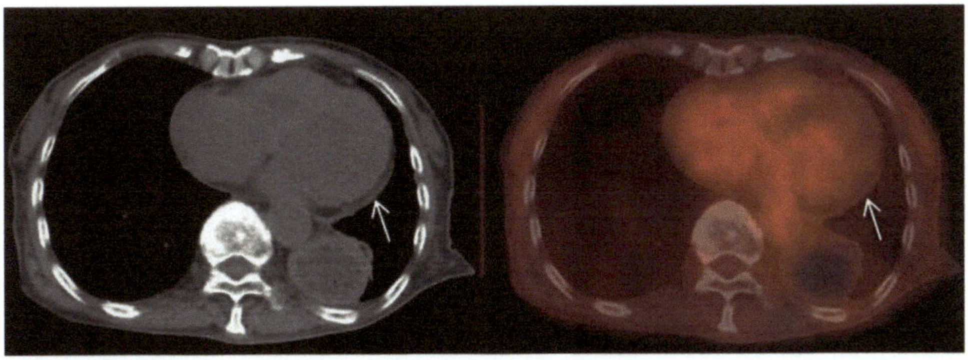

Figure 3.59 Computed tomography (left) and PET/computed tomography fusion images (right). There is mild pericardial effusion (white arrows). The CT value of the cardiac cavity is lower than that of the myocardium, indicating anemia. *PET*, Positron Emission Tomography.

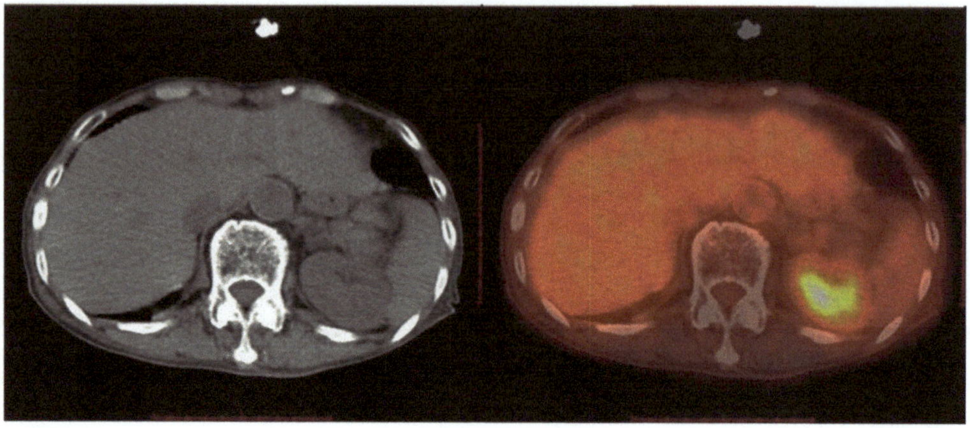

Figure 3.60 Computed tomography (left) and PET/computed tomography fusion images (right). There is a diffuse increase of the radioactive uptake in the liver. *PET*, Positron Emission Tomography.

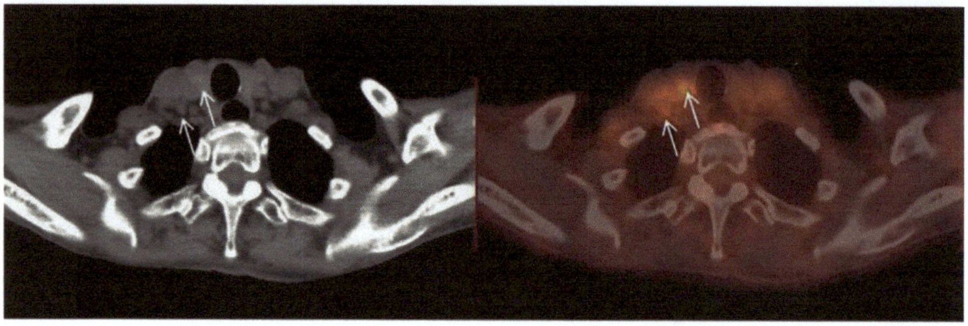

Figure 3.61 Computed tomography (left) and PET/computed tomography fusion images (right). There is a hypodense nodule in the right lobe of the thyroid gland with increased metabolic activity, with peripheral normal-sized lymph node (white arrows). Malignant nodule should be considered. *PET*, Positron Emission Tomography.

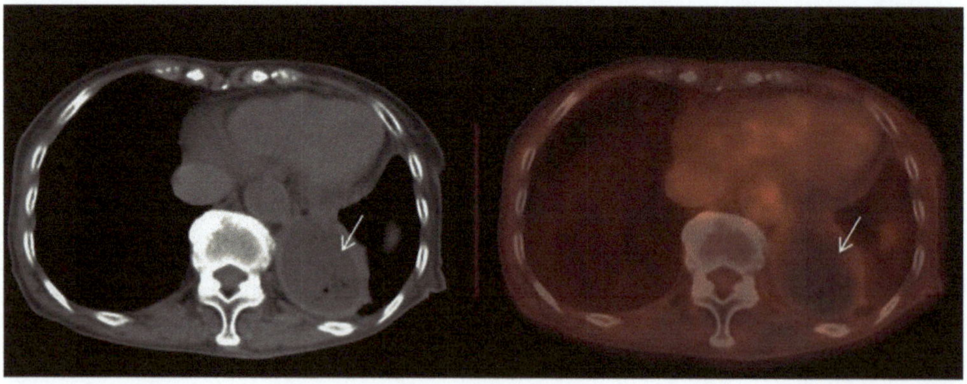

Figure 3.62 Computed tomography (left) and PET/computed tomography fusion images (right). Thoracic stomach (white arrows) is the postoperative change of esophageal cancer. *PET*, Positron Emission Tomography.

Case 45

Medical history

A 77-year-old man had cough, and expectoration with purulent sputum, about 1 month prior to the date of admission, accompanied by palpitation, chest distress, dyspnea, and poor appetite. He was diagnosed with "upper respiratory tract infection" at the local clinic in the epidemic area and was given intravenous "cephalosporin antibiotics" treatment. His appetite improved, whereas the cough did not go away. The throat swab nucleic acid test was positive 18 days prior to the admission date.

Laboratory findings

White blood cell count: 3.70×10^9/L, lymphocyte count: 0.65×10^9/L, lymphocyte percentage: 17.7%, C-reactive protein: 14.2 mg/L.

Radiological findings

The radiological findings of chest CT scans during his hospitalization are as follows: Figs. 3.63 and 3.64.

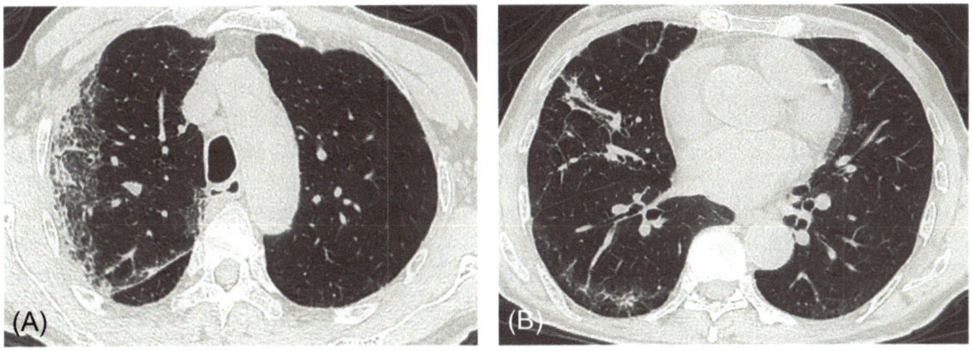

Figure 3.63 Chest computed tomography images on the day of admission. (A) There are multiple peripheral consolidations and ground-glass opacities in the right lung. The inter- and intralobular septal thickening (crazy-paving sign), parenchymal bands, and bronchiectasis are seen in the lesions. (B) Multiple ground glass opacities and fibrotic bands are noted in the right lung.

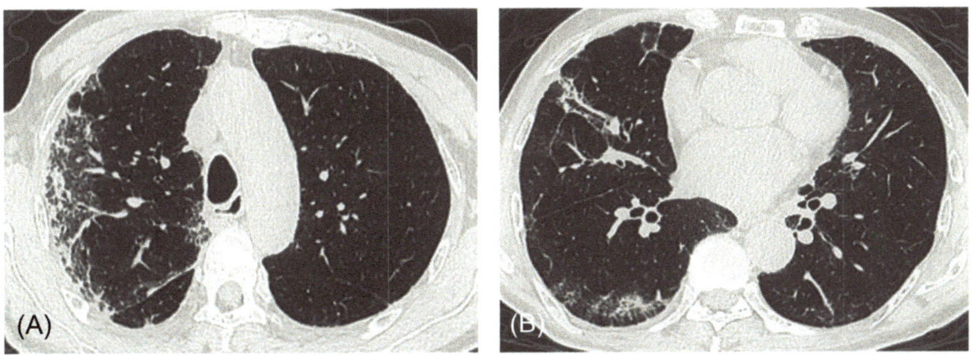

Figure 3.64 Chest computed tomography images 4 days after admission. (A) The lesions in the upper lobe of the right lung are partially absorbed. (B) The lesions in the right lung slightly enlarge.

The patient underwent ^{18}F-FDG PET/CT just 5 days after admission, and the image manifestations are as follows: Figs. 3.65–3.71.

The radiological findings of chest CT scans after discharge are as follows:

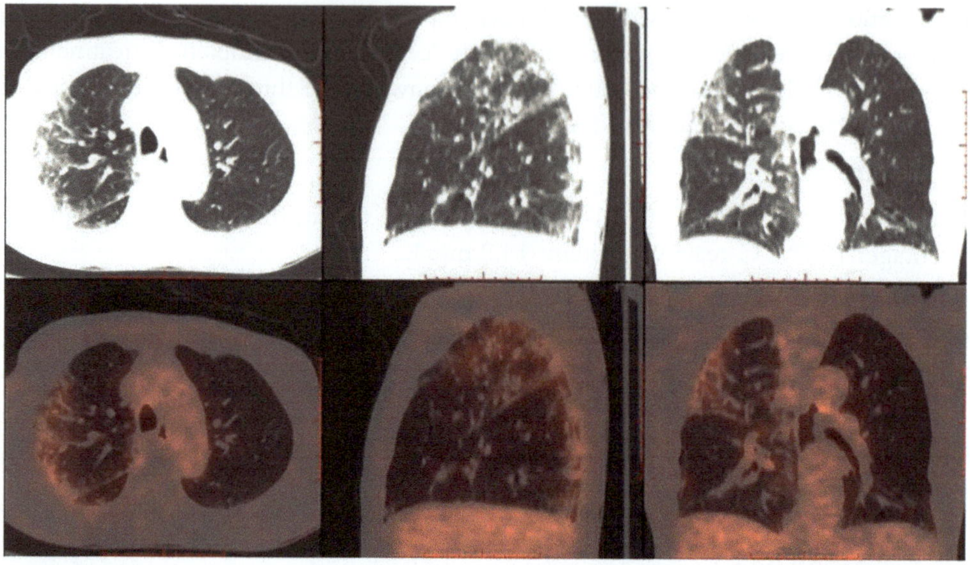

Figure 3.65 Computed tomography (top panel), PET/computed tomography fusion images (bottom panel) and 3D reconstruction images (lung window). There are multiple peripheral ground-glass and consolidative opacities, inter- and intra-lobular septal thickening, and parenchymal bands predominant in the right lung. The subpleural peripheral lesions show mild increased metabolic activity, indicating the changes in the absorption period of COVID-19, with a small part of lesions showing mild increased metabolic activity. *PET*, Positron Emission Tomography.

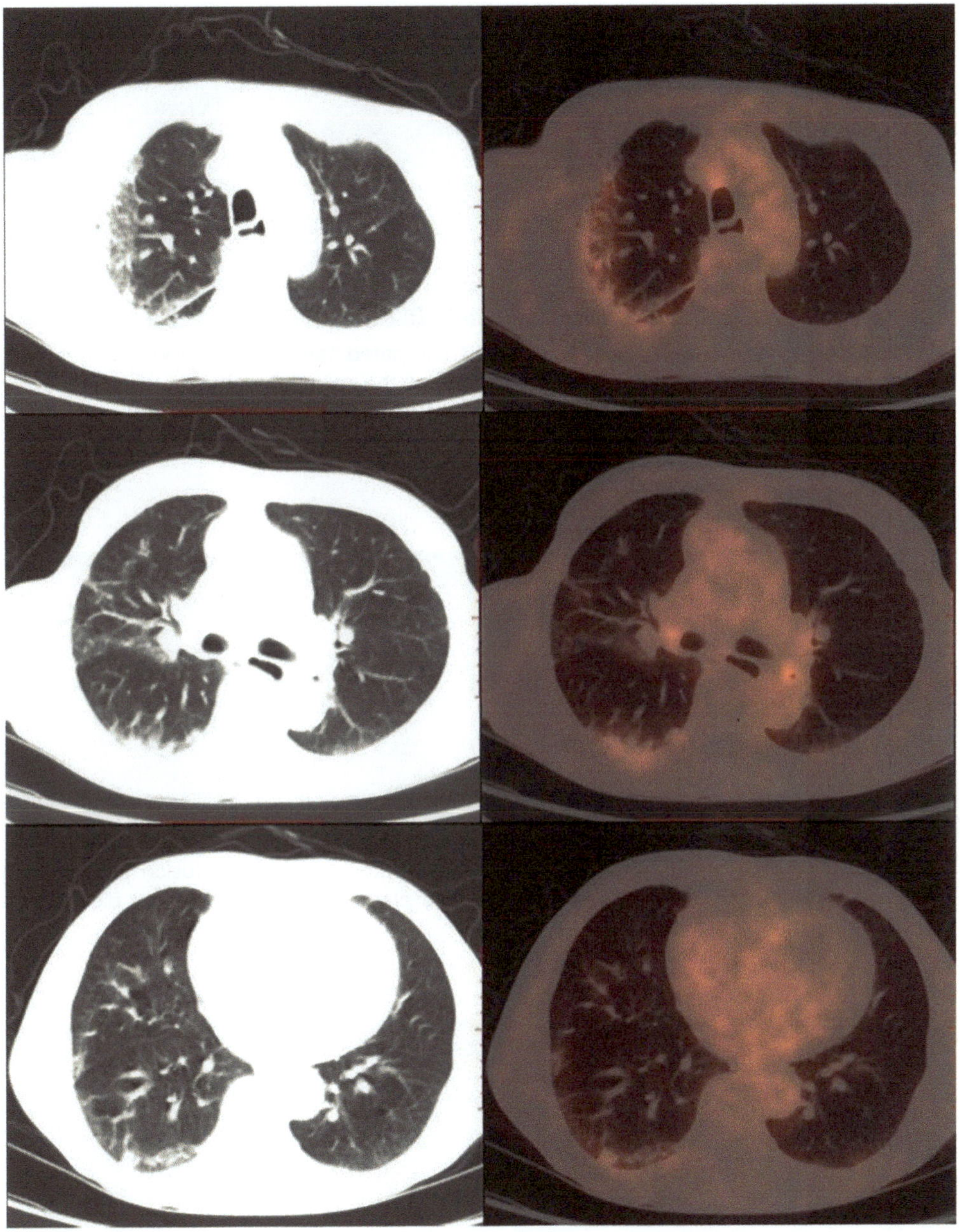

Figure 3.66 Computed tomography (left panel) and PET/computed tomography fusion images (right panel) of multiple axial slices (lung window). There are multiple peripheral ground-glass and consolidative opacities, inter- and intra-lobular septal thickening, and parenchymal bands predominant in the right lung. The subpleural peripheral lesions show mild increased metabolic activity, indicating the changes in the absorption period of COVID-19, with a small part of the lesions showing mild increased metabolic activity. *PET*, Positron Emission Tomography.

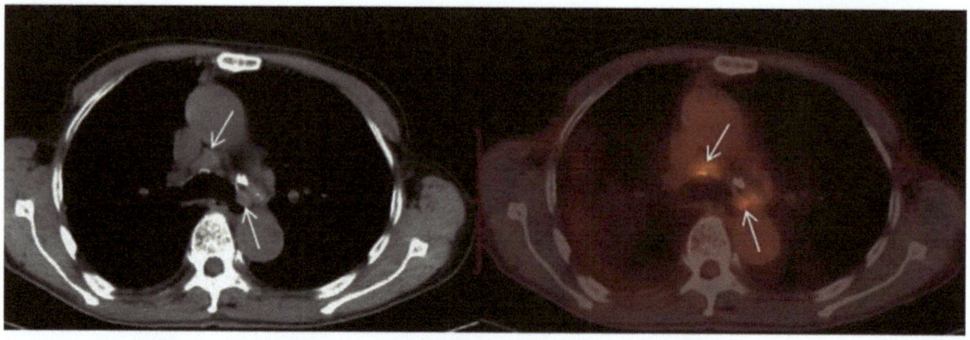

Figure 3.67 Computed tomography (left) and PET/computed tomography fusion images (right). There are multiple normal-sized lymph nodes located at the aortopulmonary level (white arrows). They show isodensity on computed tomography and increased metabolic activity, suggesting a diagnosis of inflammatory lymph nodes. *PET*, Positron Emission Tomography.

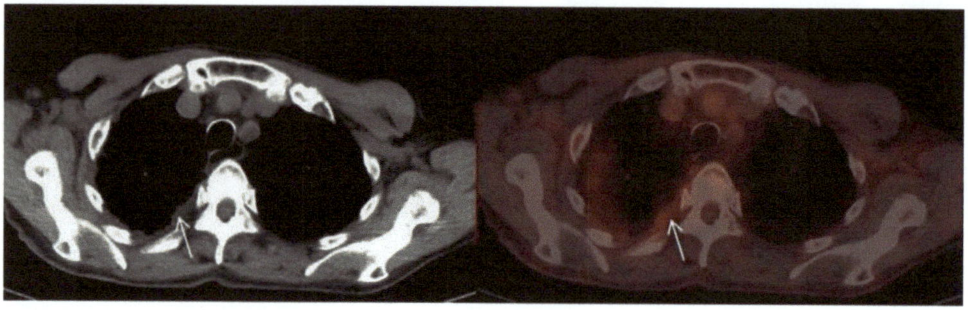

Figure 3.68 Computed tomography (left) and PET/computed tomography fusion images (right). The right pleura is slightly thickened, and there is mild pleural effusion in the right pleural cavity. *PET*, Positron Emission Tomography.

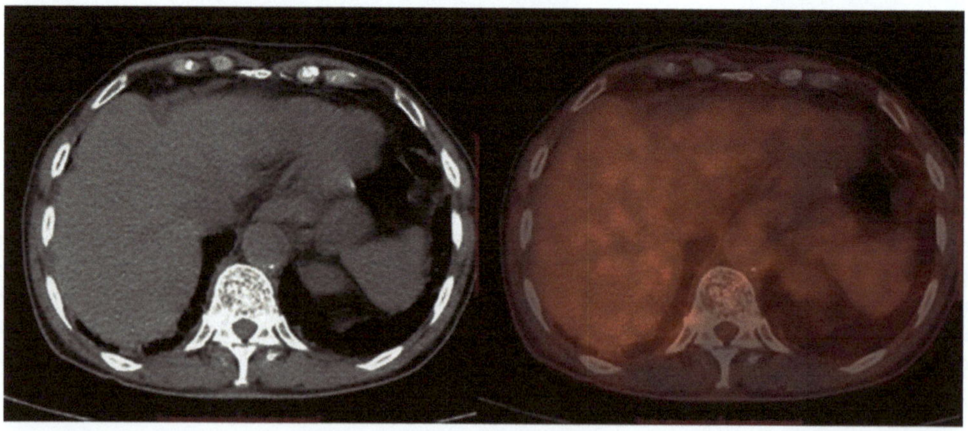

Figure 3.69 Computed tomography (left) and PET/computed tomography fusion images (right). The spleen is small-sized. *PET*, Positron Emission Tomography.

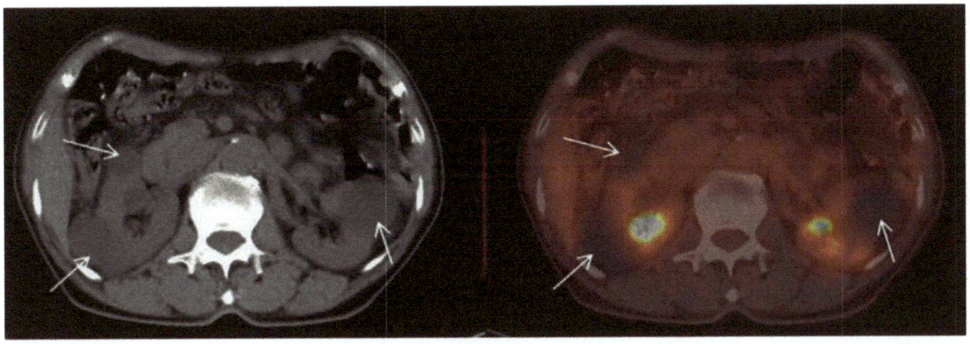

Figure 3.70 Computed tomography (left) and PET/computed tomography fusion images (right). There are multiple cysts in both kidneys. *PET*, Positron Emission Tomography.

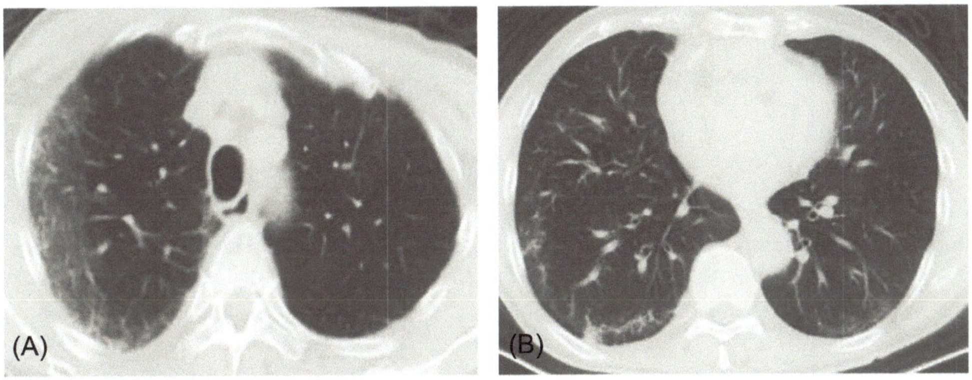

Figure 3.71 Chest computed tomography images 15 days after discharge. (A) The lesions in the upper lobe of the right lung are partially absorbed. (B) The lesions in both lungs are partially absorbed.

Case 46

Medical history

The patient is a 79-year-old woman. She documented a history of contact with people diagnosed with COVID-19. She reported fever 18 days prior to the date of admission with a maximum temperature of 38.5°C, accompanied by a paroxysmal dry cough. The nucleic acid test was positive 14 days prior to the admission date. After symptomatic treatment, the body temperature returned to normal while cough still persisted. She developed worsened symptoms and chest tightness 16 days prior to the date of admission. Blood biochemical examination showed the potassium concentration was 2.62 mmol/L on the day of admission.

Laboratory findings

White blood cell count: $7.60 \times 10^9/L$, lymphocyte count: $1.04 \times 10^9/L$, lymphocyte percentage: 13.7%, C-reactive protein: 10.9 mg/L.

Radiological findings

The radiological findings of chest CT scans during her hospitalization are as follows: Figs. 3.72 and 3.73

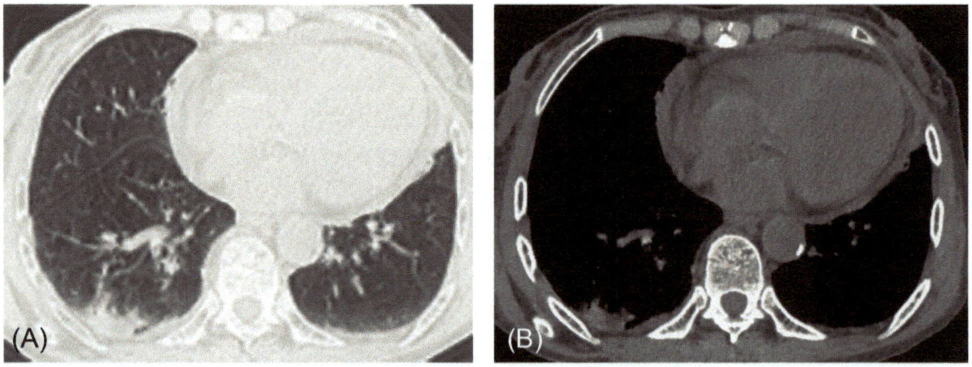

Figure 3.72 Chest computed tomography images on the day of admission. (A) Consolidations in the right lower lobe are noted. (B) Bilateral pleural effusions and pericardial effusions are noted.

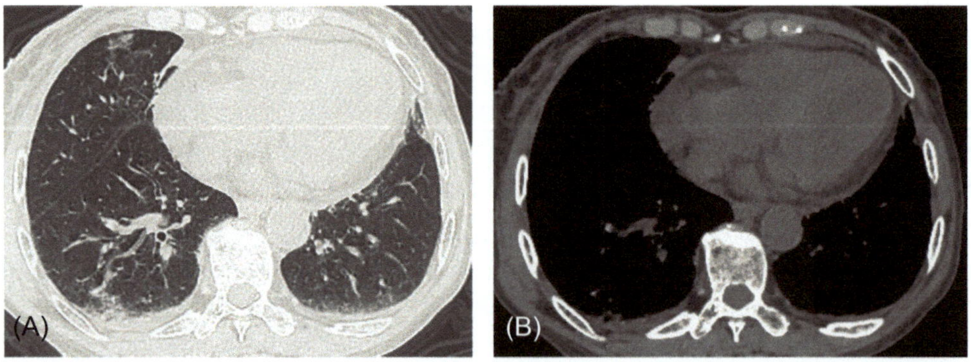

Figure 3.73 Chest computed tomography images 4 days after admission. (A) There is a reduction in the area of lesions in the right lower lobe, and there is a new emerging ground-glass opacity in the right middle lobe. (B) Minimal bilateral pleural effusions and pericardial effusions are noted.

The patient underwent ^{18}F-FDG PET/CT just 5 days after admission, and the image manifestations are as follows: Fig. 3.74.

The subpleural lesions show mild increased metabolic activity, indicating changes in the absorption period of COVID-19, with a small part of the lesions showing mild increased metabolic activity (Figs. 3.75–3.79).

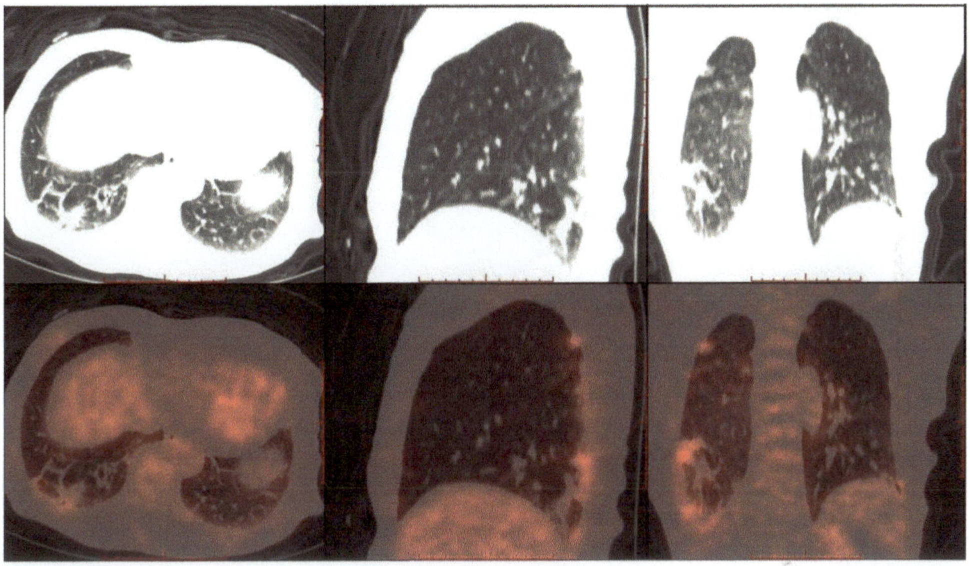

Figure 3.74 Computed tomography (top panel), PET/computed tomography fusion images (bottom panel) and 3D reconstruction images (lung window). There are multiple consolidative opacities and parenchymal bands. *PET*, Positron Emission Tomography.

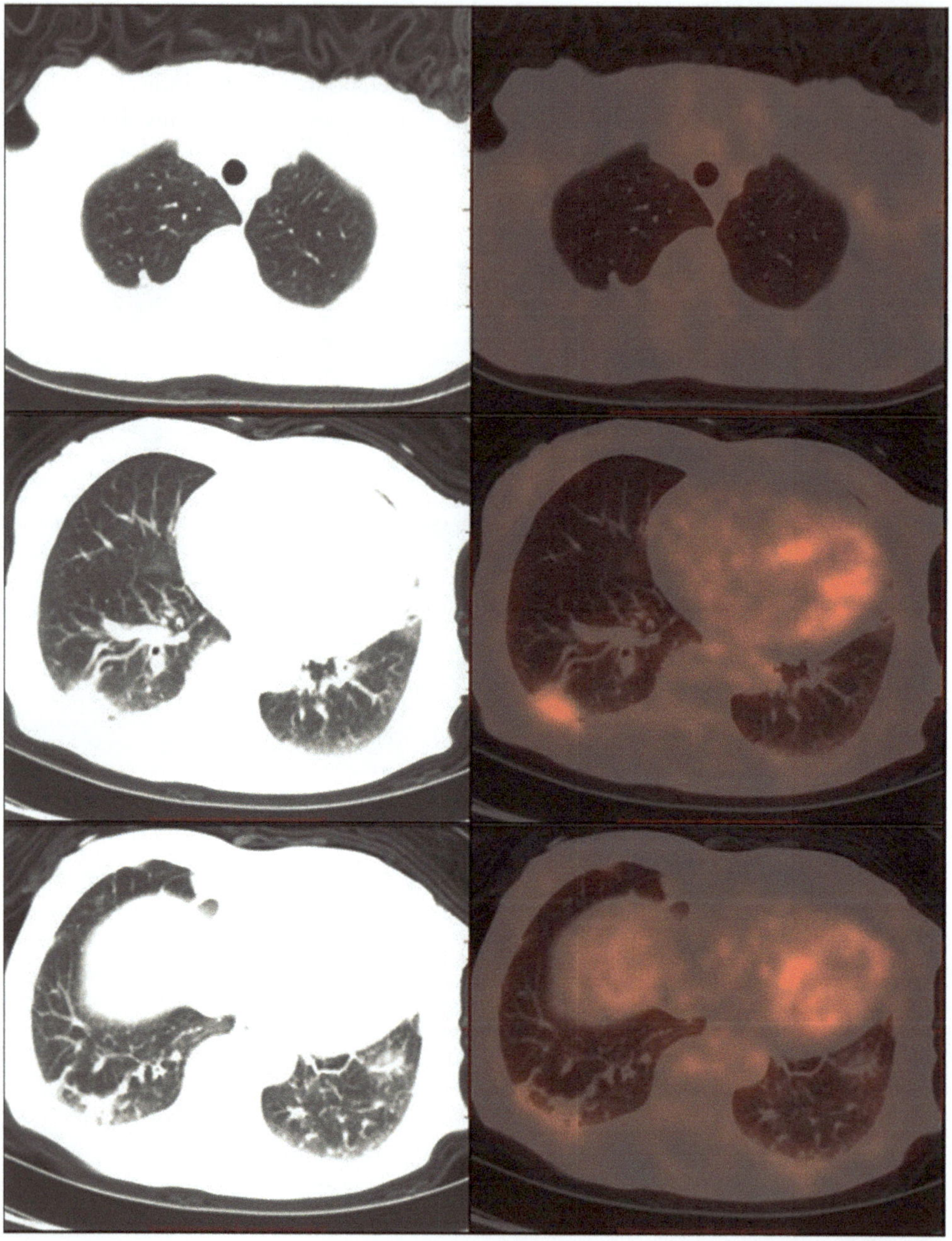

Figure 3.75 Computed tomography (left panel) and PET/computed tomography fusion images (right panel) of multiple axial slices (lung window). There are multiple consolidative opacities and parenchymal bands. The subpleural lesions show mild increased metabolic activity, indicating changes in the absorption period of COVID-19, with a small part of the lesions showing mild increased metabolic activity. *PET*, Positron Emission Tomography.

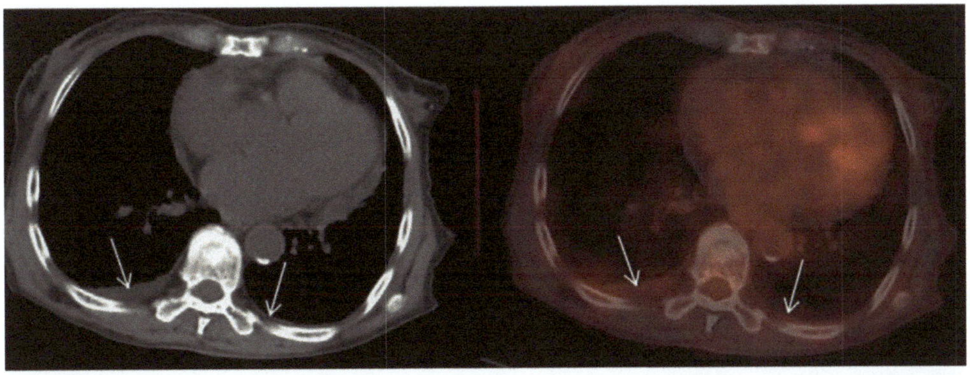

Figure 3.76 Computed tomography (left) and PET/computed tomography fusion images (right). There is mild pleural effusion on both sides (white arrows). *PET*, Positron Emission Tomography.

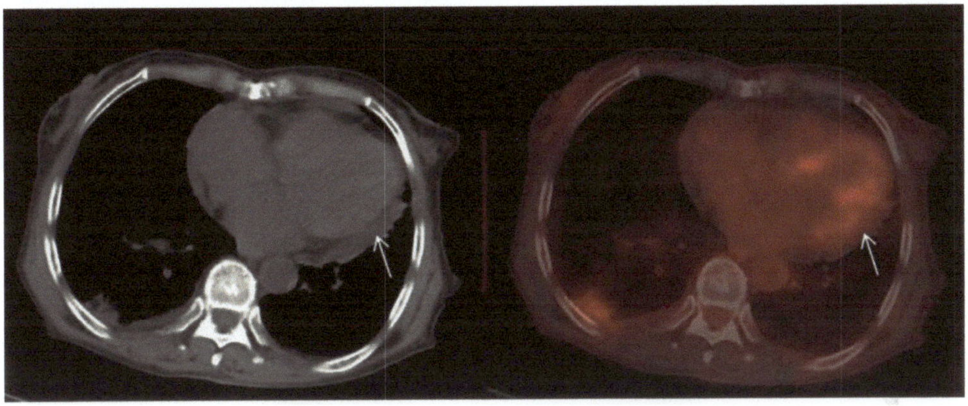

Figure 3.77 Computed tomography (left) and PET/computed tomography fusion images (right). There is mild pericardial effusion (white arrows). *PET*, Positron Emission Tomography.

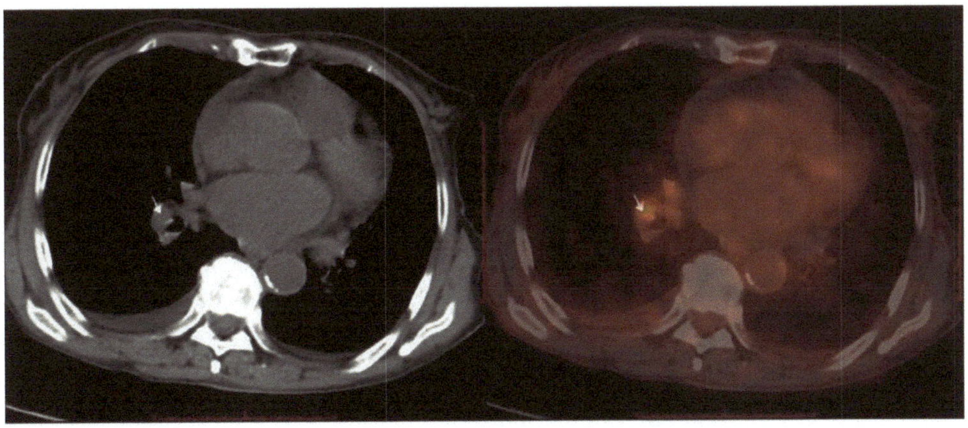

Figure 3.78 Computed tomography (left) and PET/computed tomography fusion images (right). There are multiple normal-sized lymph nodes located in both pulmonary hilum area (white arrows). They show isodensity on computed tomography and hypermetabolism on PET, suggesting a diagnosis of inflammatory lymph nodes. *PET*, Positron Emission Tomography.

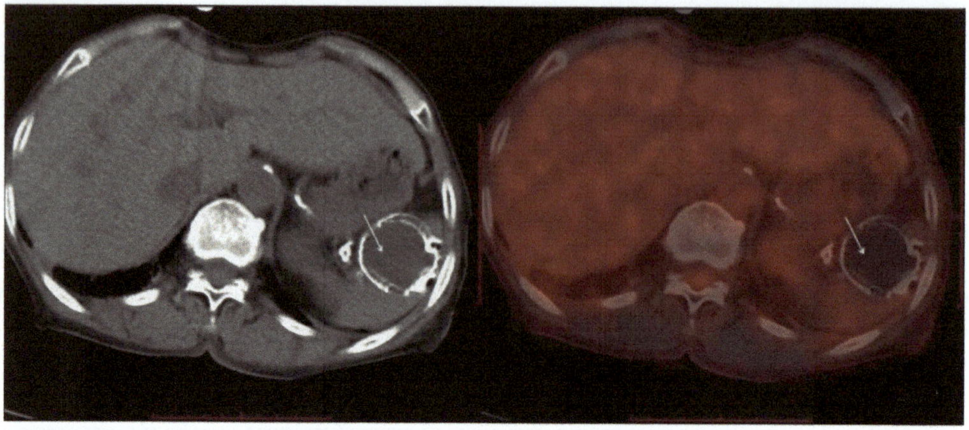

Figure 3.79 Computed tomography (left) and PET/computed tomography fusion images (right). There is a circular calcified cystic benign lesion in the spleen. *PET*, Positron Emission Tomography.

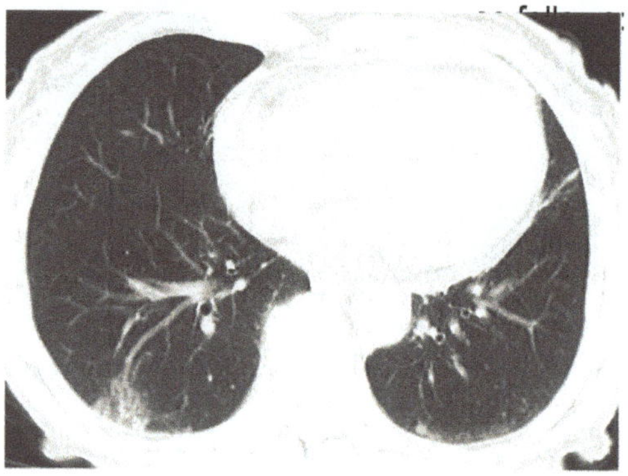

Figure 3.80 Chest computed tomography image 15 days after discharge. Axial chest computed tomography images show similar manifestations to prior images.

The radiological findings of chest CT scans after discharge are as follows: Fig. 3.80.

Case 47

Medical history

A 56-year-old woman reported fever with a maximum temperature of 38.5°C just 15 days prior to the admission, accompanied by cough, expectoration with white sticky sputum, fatigue, headache, muscle pain, chills, poor appetite, and chest distress.

She was treated as suffering from an upper respiratory tract infection by a local clinic without effect. The throat swab nucleic acid test was positive 5 days prior to the date of admission, and he started to have chest tightness 1 day prior to the admission date. Past medical history includes a hysterectomy for cervical precancerous lesions 1 year prior to the date of admission.

Laboratory findings on admission

White blood cell count: 6.10×10^9/L, lymphocyte count: 1.32×10^9/L, lymphocyte percentage: 21.5%, C-reactive protein: 6.8 mg/L.

Laboratory findings 1 year and 7 months after discharge

White blood cell count: 7.19×10^9/L, lymphocyte count: 2.78×10^9/L, lymphocyte percentage:38.7%, C-reactive protein: <0.499 mg/L.

Radiological findings

The radiological findings of chest CT scans during her hospitalization are as follows: Figs. 3.81–3.83.

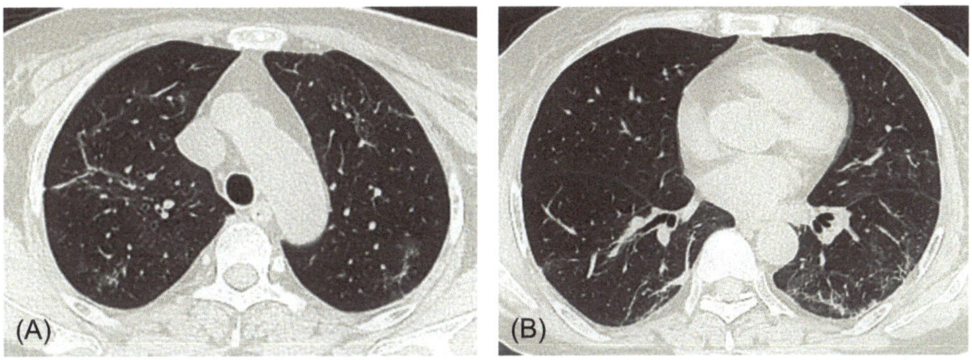

Figure 3.81 Chest computed tomography images on the day of admission. (A) Multiple ground-glass opacities and parenchymal bands are noted in bilateral lungs. (B) Multiple ground-glass opacities and parenchymal bands are noted in bilateral lungs.

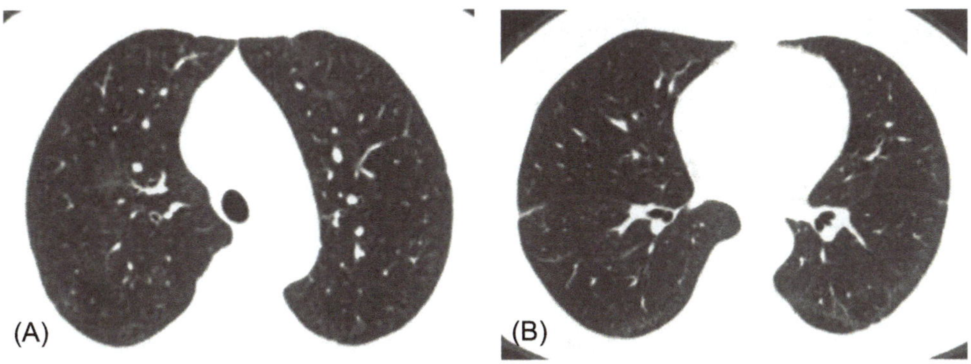

Figure 3.82 Chest computed tomography images 14 days after discharge. (A) Multiple ground-glass opacities remained in bilateral lungs. (B) Multiple ground-glass opacities remained in bilateral lungs.

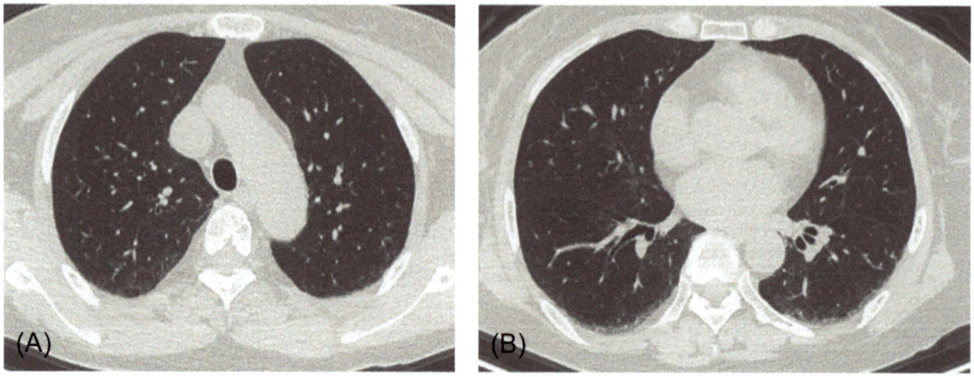

Figure 3.83 Chest computed tomography images 1 year and 7 months after discharge. (A) Hypostatic effect is noted in bilateral lungs. (B) Hypostatic effect is noted in bilateral lungs.

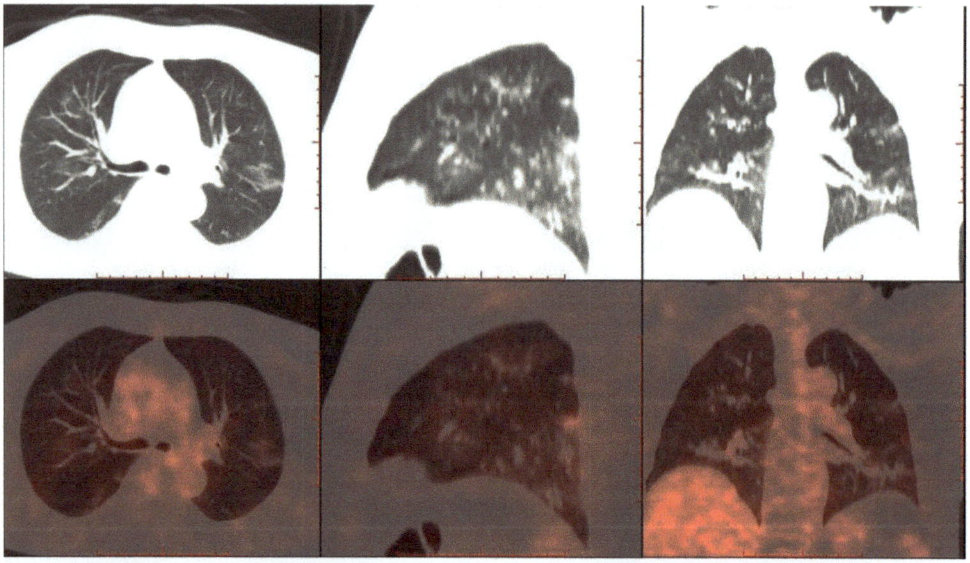

Figure 3.84 Computed tomography (top panel), PET/computed tomography fusion images (bottom panel) and 3D reconstruction images (lung window) 5 days after admission. There are multiple ground-glass and consolidative opacities in both lungs. The ground-glass opacities do not show increased metabolic activity, and some of the consolidative opacities show increased metabolic activity, suggesting a diagnosis of the absorption period of COVID-19, with a part of the lesions showing increased metabolic activity. *PET*, Positron Emission Tomography.

The patient underwent ^{18}F-FDG PET/CT, and the image manifestations are as follows: Figs. 3.84–3.89.

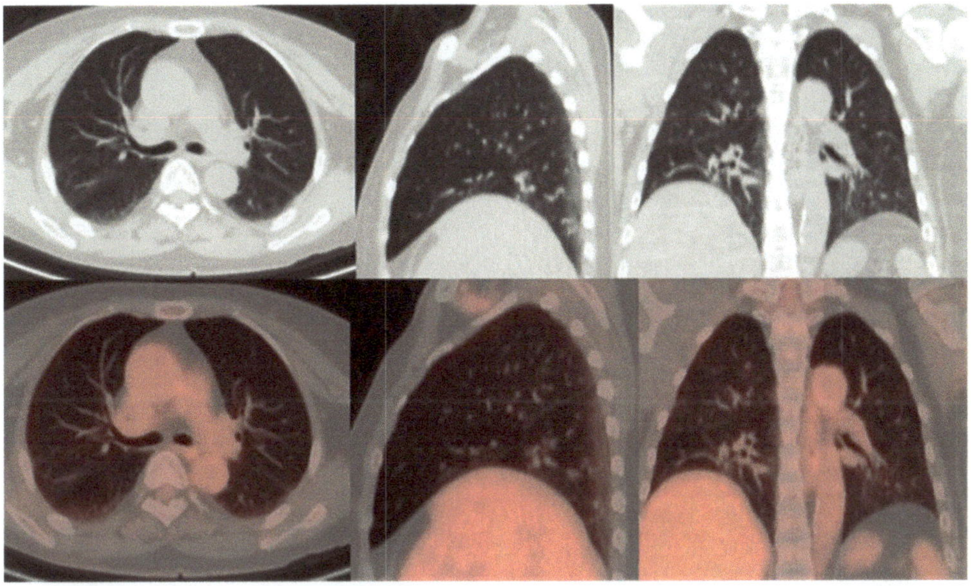

Figure 3.85 Computed tomography (top panel), PET/computed tomography fusion images (bottom panel) and 3D reconstruction images (lung window) 1 year and 7 months after discharge. The previous multiple ground-glass and consolidative opacities disappeared in both lungs. There is no area with increased metabolic activity in either lung. *PET*, Positron Emission Tomography.

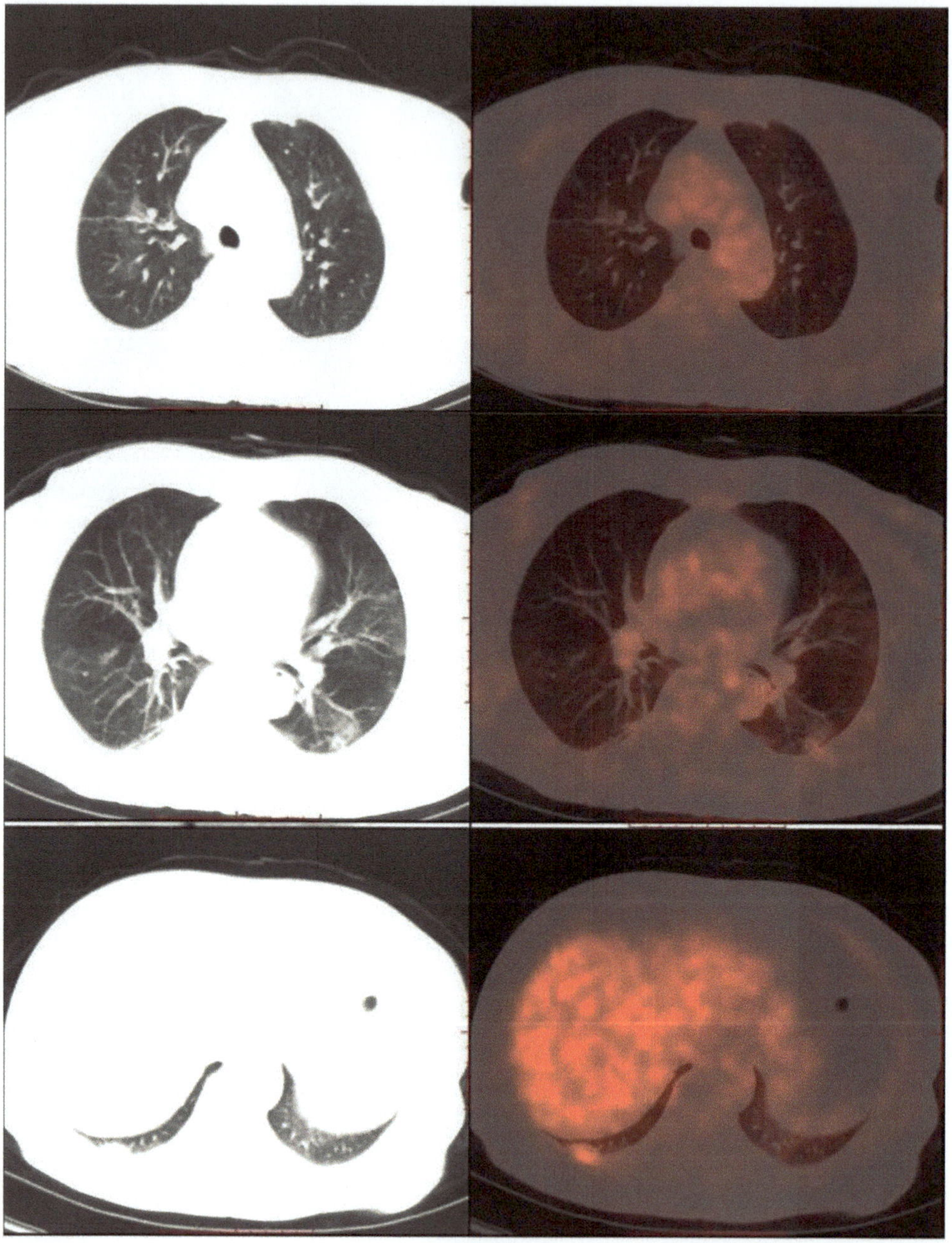

Figure 3.86 Computed tomography (left panel) and PET/computed tomography fusion images (right panel) of multiple axial slices (lung window). There are multiple ground-glass and consolidative opacities in both lungs. The ground-glass opacities do not show increased metabolic activity, and some of the consolidative opacities show increased metabolic activity, suggesting a diagnosis of the absorption period of COVID-19, with a part of the lesions showing increased metabolic activity. *PET*, Positron Emission Tomography.

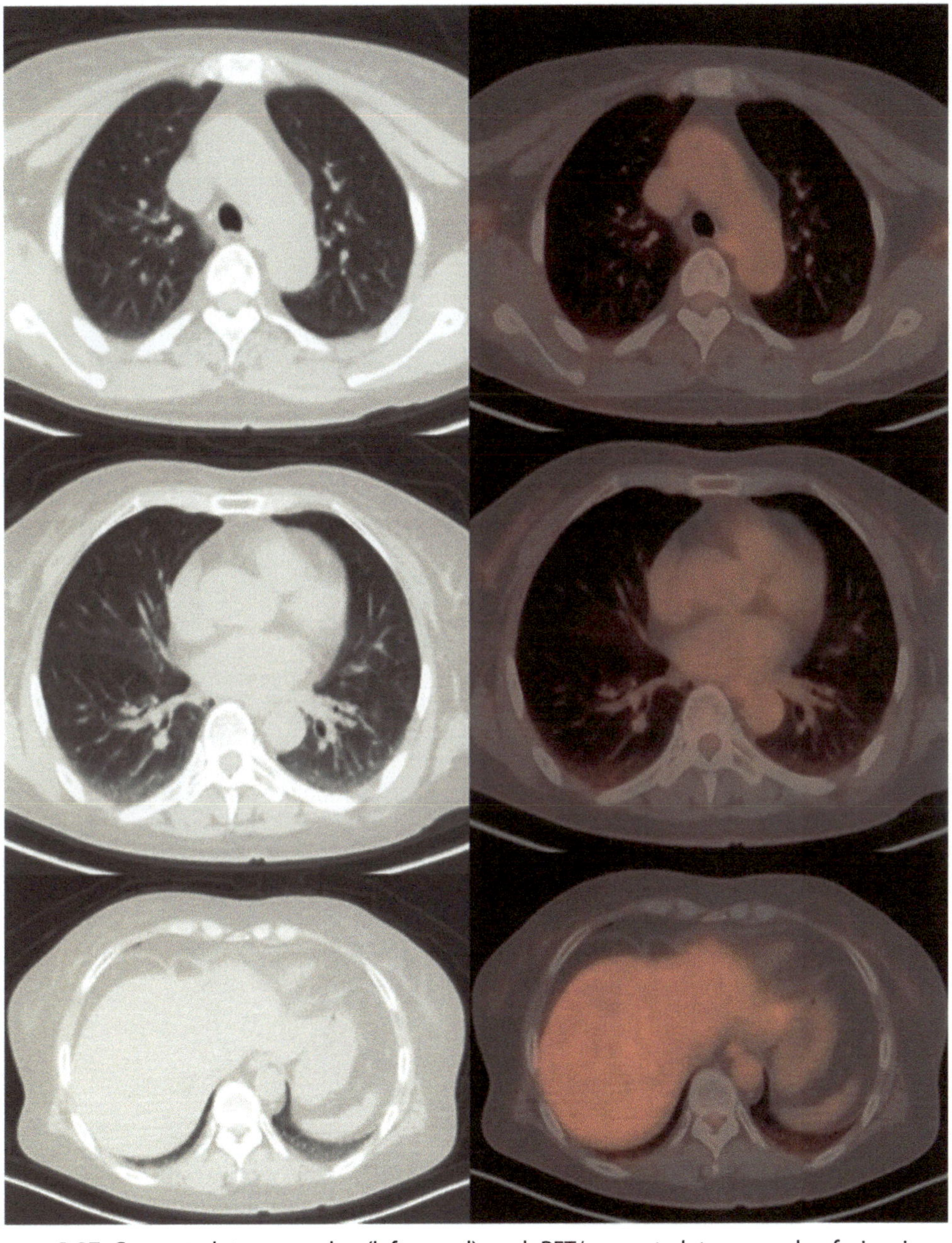

Figure 3.87 Computed tomography (left panel) and PET/computed tomography fusion images (right panel) of multiple axial slices (lung window) 1 year and 7 months after discharge. The previous multiple ground-glass and consolidative opacities have disappeared in both lungs. There is no area with increased metabolic activity in either lung. *PET*, Positron Emission Tomography.

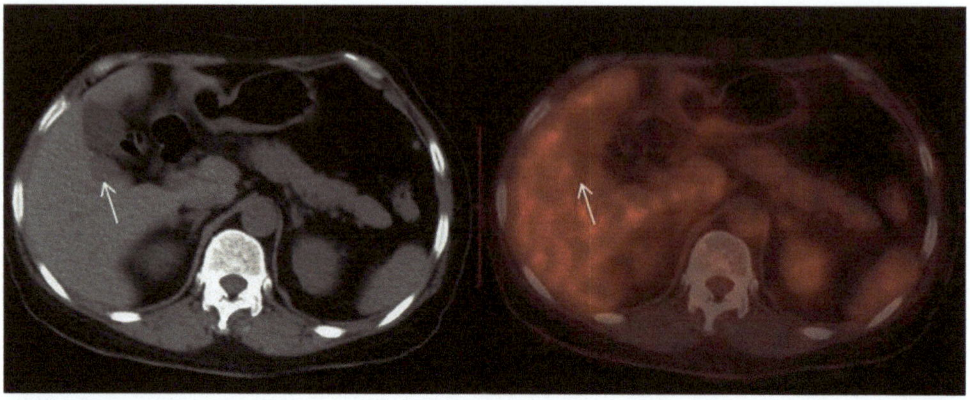

Figure 3.88 Computed tomography (left), PET/computed tomography fusion images (right) 5 days after admission. There is one hypodense lesion in the liver with hypometabolic activity (white arrows). *PET*, Positron Emission Tomography.

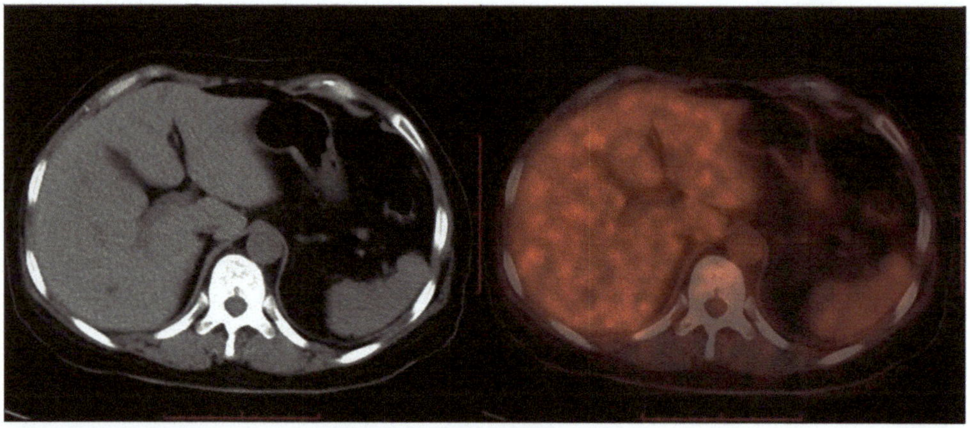

Figure 3.89 Computed tomography (left), PET/computed tomography fusion images (right) 5 days after admission. The spleen is small-sized. *PET*, Positron Emission Tomography.

Case 48

Medical history

A 57-year-old man reported fever 26 days prior to the date of admission, with a maximum temperature of 38.5°C, accompanied by chills, poor appetite, fatigue, headache, and back pain. The nucleic acid test was positive 8 days prior to the admission.

Laboratory findings on admission

White blood cell count: 3.20×10^9/L, lymphocyte count: 1.11×10^9/L, lymphocyte percentage: 35.2%, hemoglobin: 111.5 g/L, C-reactive protein: 7.3 mg/L.

Laboratory findings 1 year and 7 months after discharge

White blood cell count: 4.66×10^9/L, lymphocyte count: 1.77×10^9/L, lymphocyte percentage: 38.0%, hemoglobin: 153.0 g/L, C-reactive protein: <0.499 mg/L.

Radiological findings

The radiological findings of chest CT scans during his hospitalization are as follows: Fig. 3.90

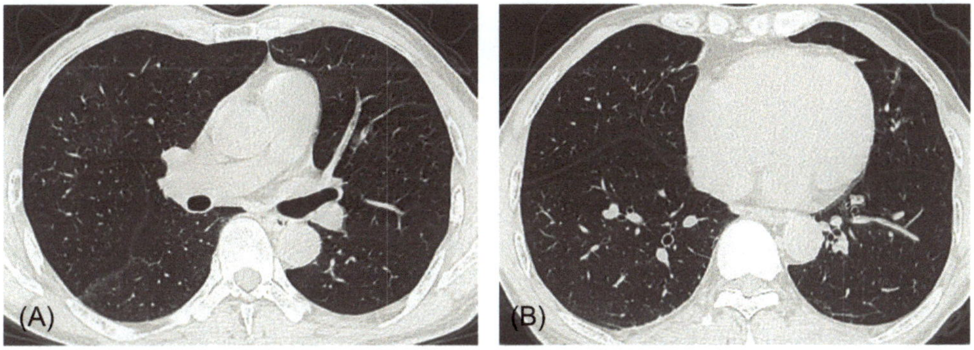

Figure 3.90 Chest computed tomography images on the day of admission. (A) There are ground-glass opacities in the right lung. (B) Multiple subpleural bands in both lungs are noted.

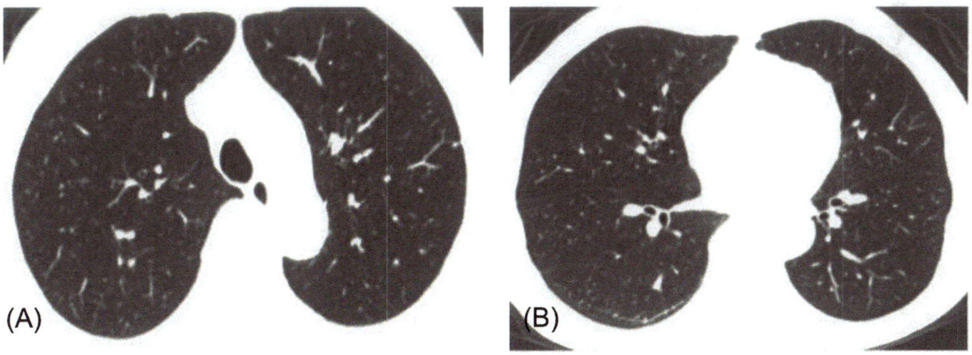

Figure 3.91 Chest computed tomography images 14 days after discharge. (A) Axial CT image shows complete disappearance of preexisting lesions. (B) Subpleural band in the right lower lobe is noted.

The radiological findings of chest CT scans after discharge are as follows: Fig. 3.91.

The patient underwent ^{18}F-FDG PET/CT, and the image manifestations are as follows: Figs. 3.92–3.98.

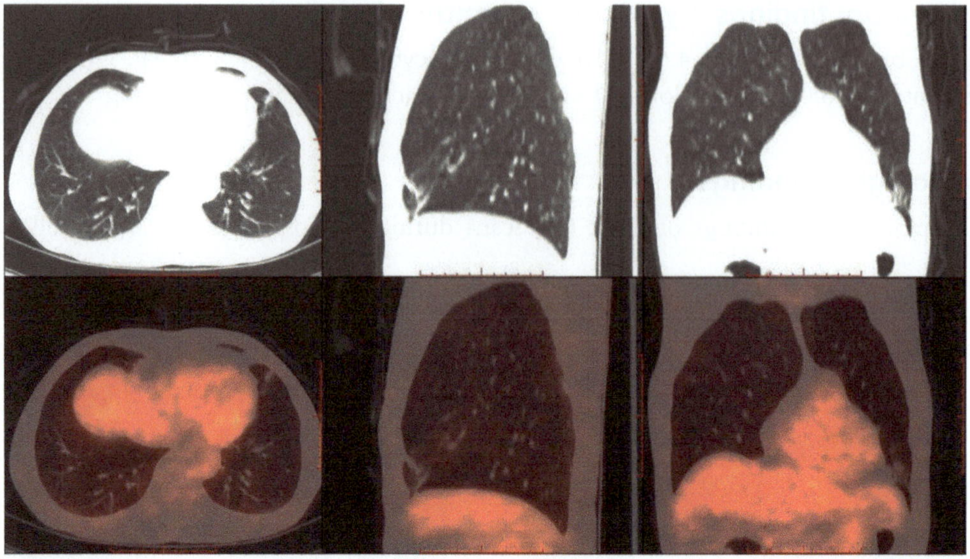

Figure 3.92 Computed tomography (top panel), PET/computed tomography fusion images (bottom panel) and 3D reconstruction images (lung window) 5 days after admission. There are mild focal ground-glass opacities and bands in both lungs without increased metabolic activity, suggesting a diagnosis of the absorption period of COVID-19, with the residual lesions showing normal metabolic activity. *PET*, Positron Emission Tomography.

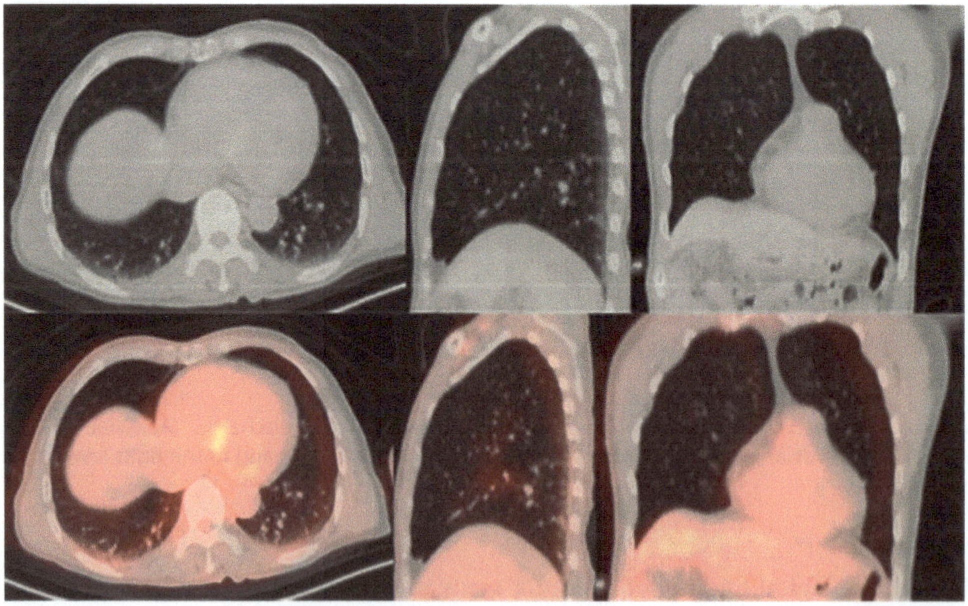

Figure 3.93 Computed tomography (top panel), PET/computed tomography fusion images (bottom panel) and 3D reconstruction images (lung window) 1 year and 7 months after discharge. The previous mild focal ground-glass opacities and bands have disappeared. *PET*, Positron Emission Tomography.

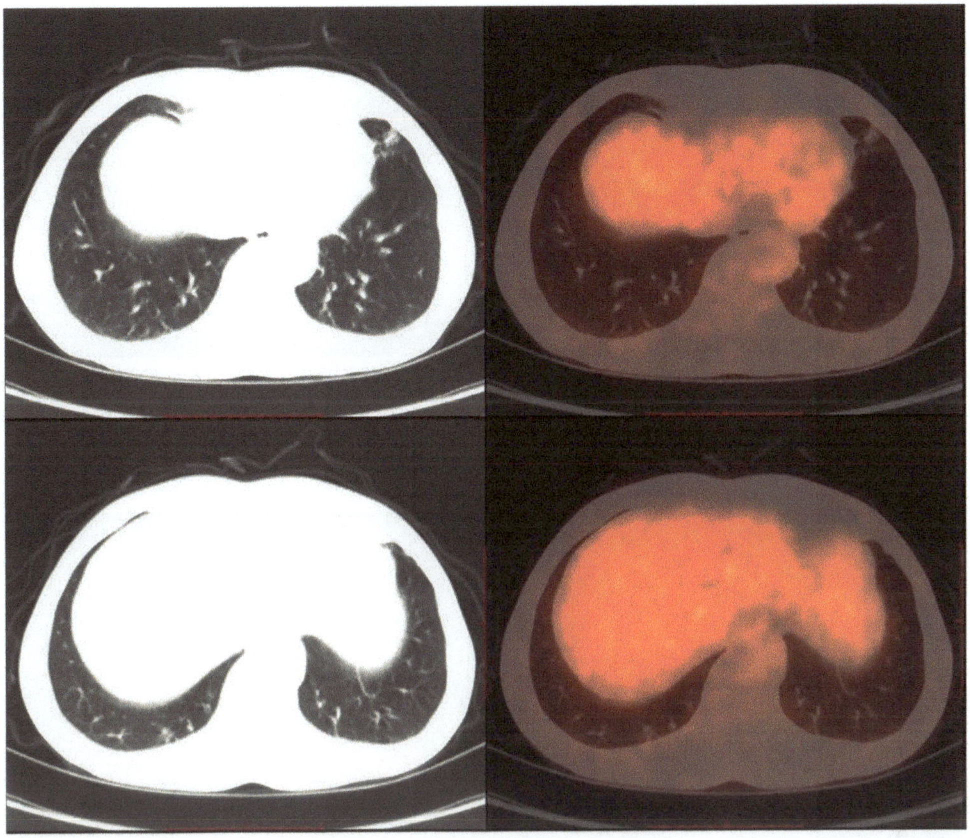

Figure 3.94 Computed tomography (left panel), PET/computed tomography fusion images (right panel) of multiple axial slices (lung window) 5 days after admission. There are mild focal ground-glass opacities and bands in both lungs without increased metabolic activity, suggesting the diagnosis of the absorption period of COVID-19, with the residual lesions showing normal metabolic activity. *PET*, Positron Emission Tomography.

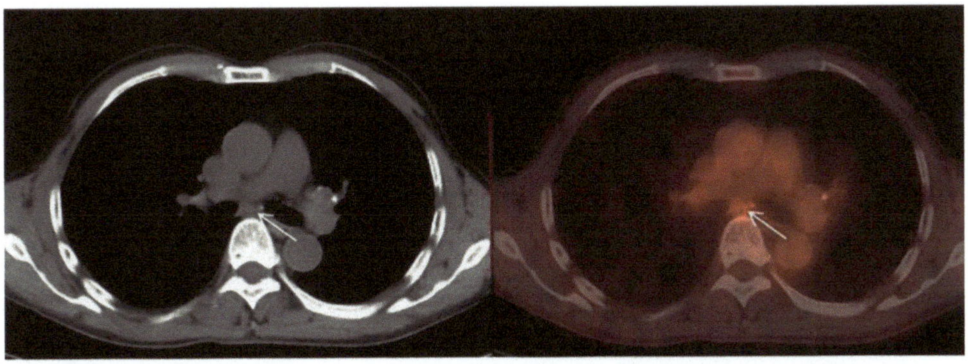

Figure 3.95 Computed tomography (left) and PET/computed tomography fusion images (right) 5 days after admission. There are normal-sized lymph nodes in the mediastinum (white arrows). They show increased metabolic activity on PET, suggesting the diagnosis of inflammatory lymph nodes. *PET*, Positron Emission Tomography.

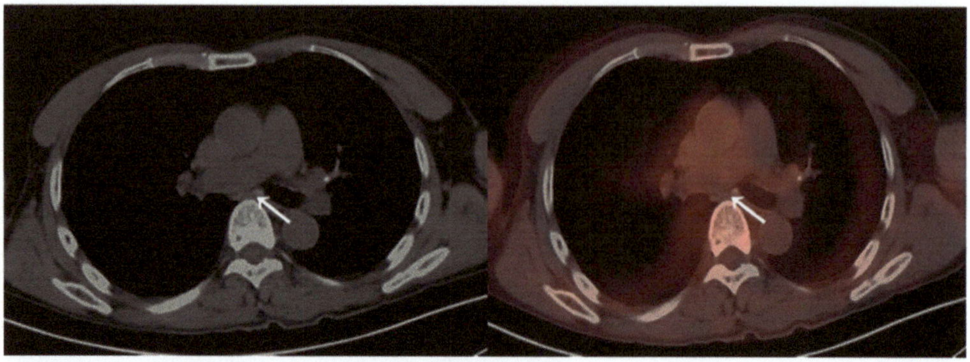

Figure 3.96 Computed tomography (left) and PET/computed tomography fusion images (right). There are normal-sized lymph nodes in the mediastinum (white arrows) 1 year and 7 months after discharge. They show normal metabolic activity on PET. *PET*, Positron Emission Tomography.

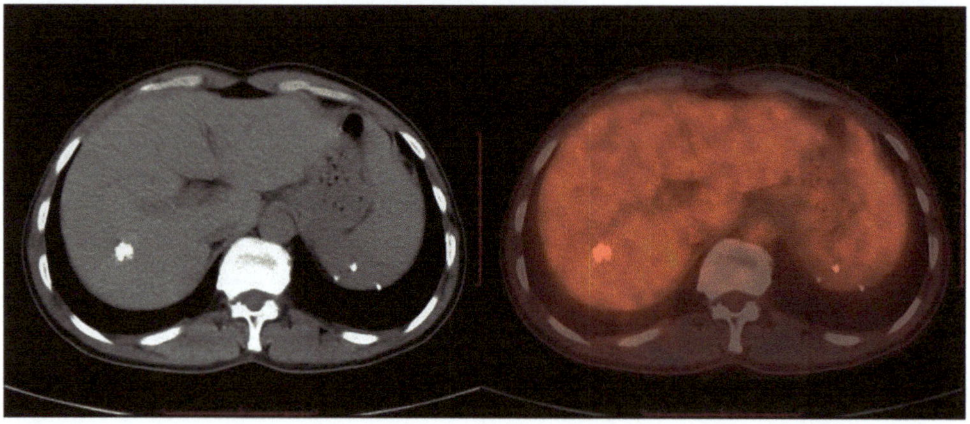

Figure 3.97 Computed tomography (left) and PET/computed tomography fusion images (right) 5 days after admission. There is diffusedly increased radiotracer uptake in the liver and multiple calcifications both in the liver and spleen. *PET*, Positron Emission Tomography.

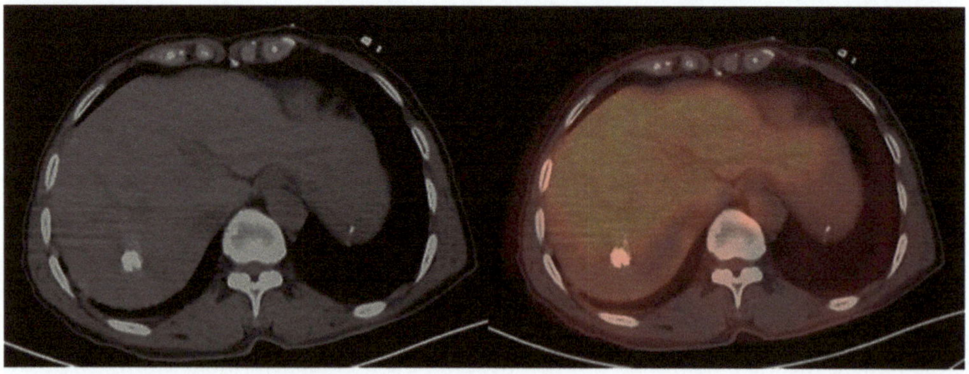

Figure 3.98 Computed tomography (left) and PET/computed tomography fusion images (right) 1 year and 7 months after discharge. The diffused increase of the radiotracer uptake is decreased in the liver. *PET*, Positron Emission Tomography.

The patient underwent PET/MR, the image manifestations are as follows: Figs. 3.99–3.107.

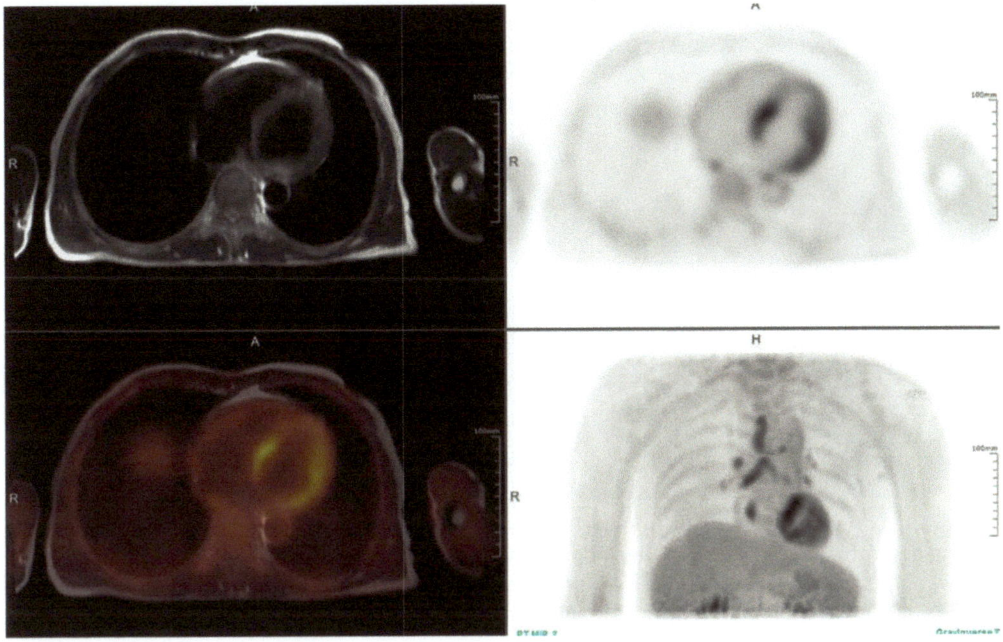

Figure 3.99 Cardiac axial black blood sequence (left upper), PET image (right upper), fusion image (left bottom), and MIP image (right bottom) 5 days after admission. There is no remarkable change in the anatomy and metabolism of the heart. *MIP*, Maximum Intensity Projection; *PET*, Positron Emission Tomography.

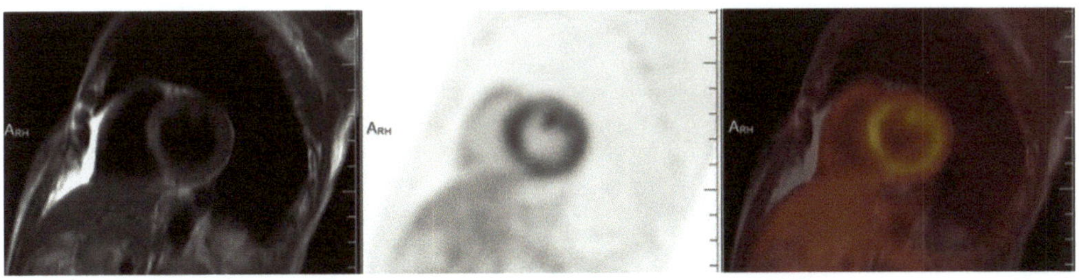

Figure 3.100 Cardiac sagittal black blood sequence (left), PET image (middle), and fusion image (right) 5 days after admission. There is no remarkable change in the anatomy and metabolism of the heart. *PET*, Positron Emission Tomography.

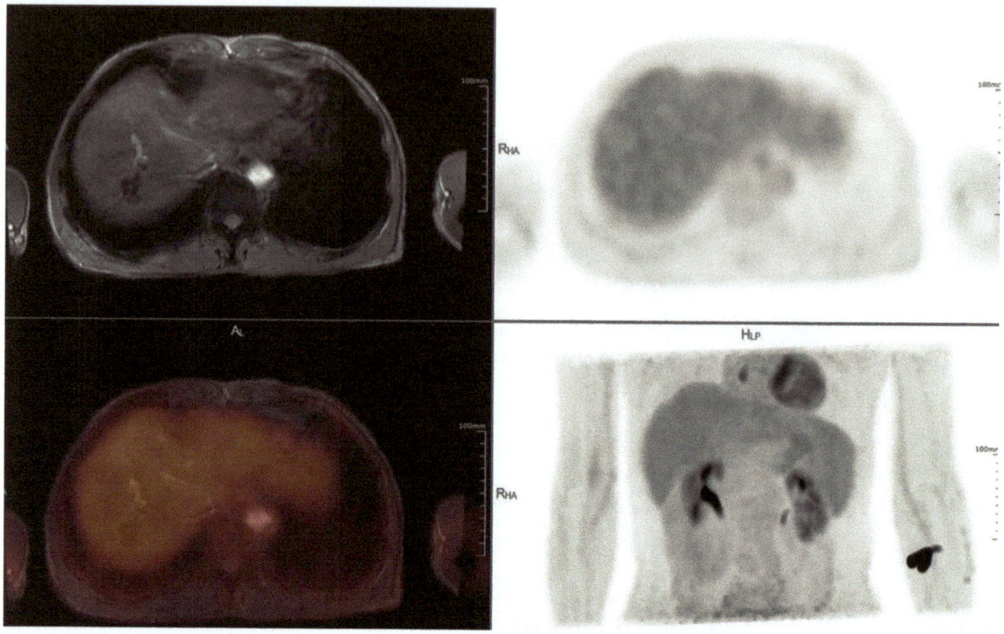

Figure 3.101 Liver axial SWI (left upper), PET image (right upper), fusion image (left bottom), and MIP image (right bottom) 5 days after admission. There is a focal calcification in the right lobe of the liver. The whole liver shows diffused increased metabolism. *MIP*, Maximum Intensity Projection; *PET*, Positron Emission Tomography; *SWI*, Susceptibility Weighted Imaging.

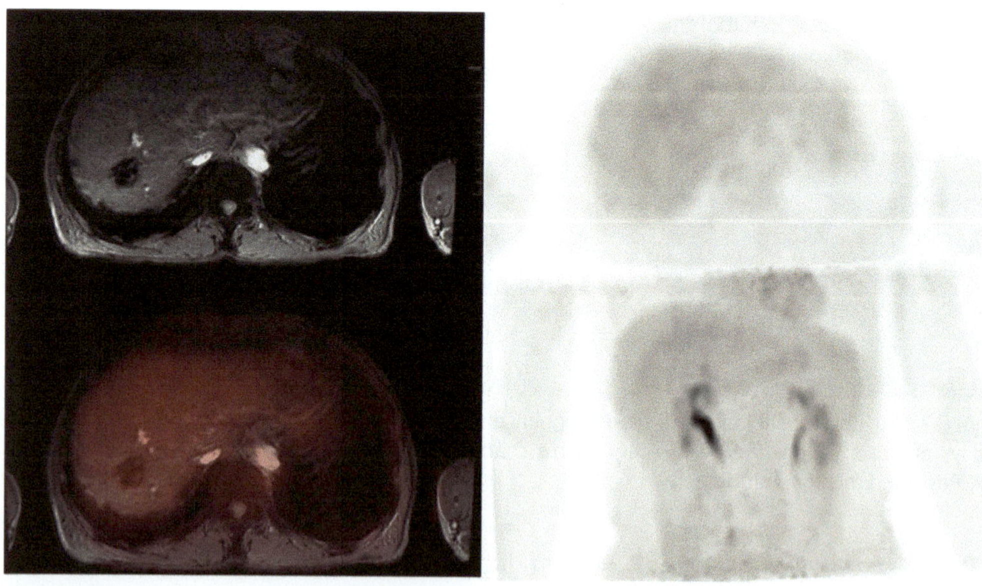

Figure 3.102 Liver axial SWI (left upper), PET image (right upper), fusion image (left bottom), and MIP image (right bottom) 1 year and 7 months after discharge. There is a focal calcification in the right lobe of the liver. The diffused increase of the radiotracer uptake is decreased in the liver. *MIP*, Maximum Intensity Projection; *PET*, Positron Emission Tomography; *SWI*, Susceptibility Weighted Imaging.

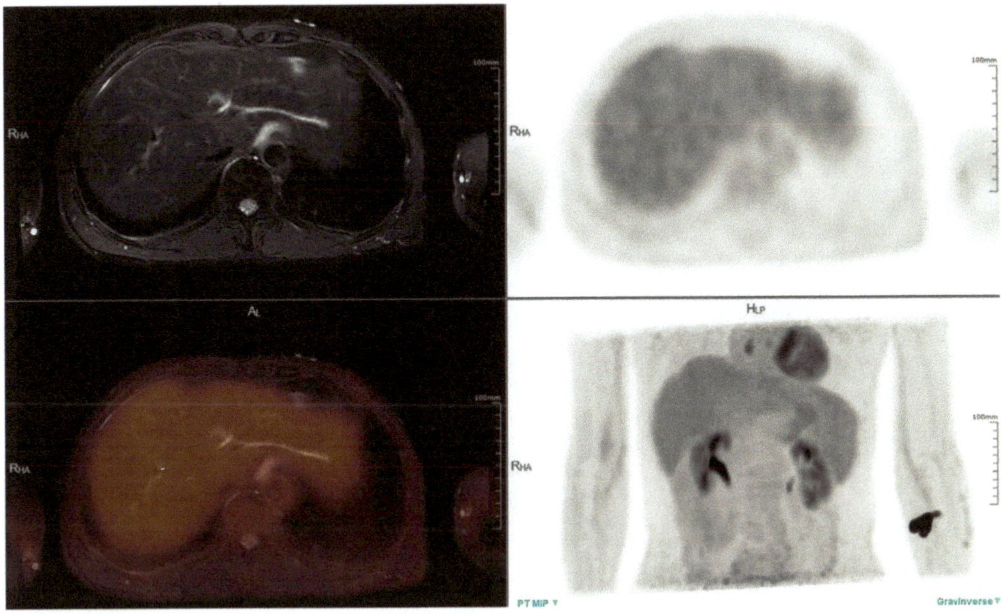

Figure 3.103 Liver axial T2WI with fat suppression (left upper), PET image (right upper), fusion image (left bottom), and MIP image (right bottom). There is a focal calcification in the right lobe of the liver. The whole liver shows diffuse increased metabolism. *MIP*, Maximum Intensity Projection; *PET*, Positron Emission Tomography; *T2WI*, T2-Weighted Imaging.

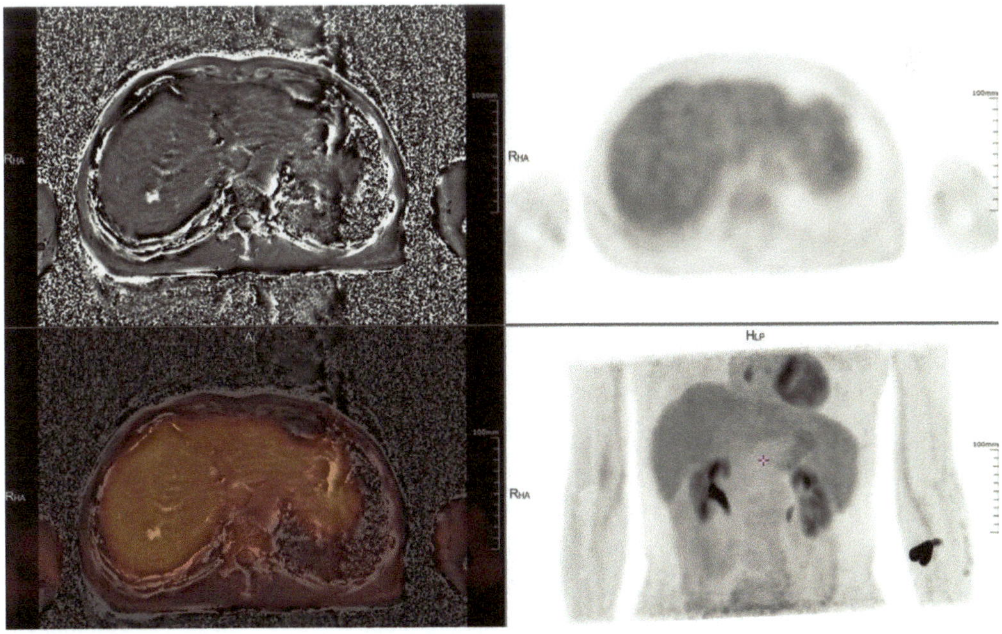

Figure 3.104 Liver axial phase image (left upper), PET image (right upper), fusion image (left bottom), and MIP image (right bottom). There is a focal calcification in the right lobe of the liver. The whole liver shows diffuse increased metabolism. *MIP*, Maximum Intensity Projection; *PET*, Positron Emission Tomography.

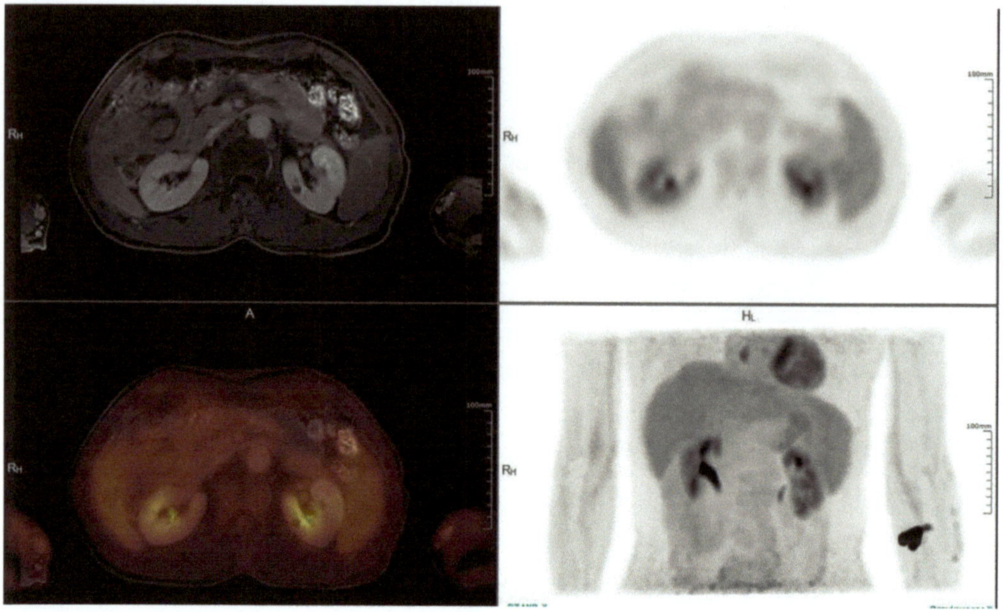

Figure 3.105 Renal axial T1WI with contrast (left upper), PET image (right upper), fusion image (left bottom), and MIP image (right bottom). There is a round hypointensity signal on enhanced T1WI with hypometabolism on PET located in the left kidney, indicating a left renal cyst. *MIP*, Maximum Intensity Projection; *PET*, Positron Emission Tomography.

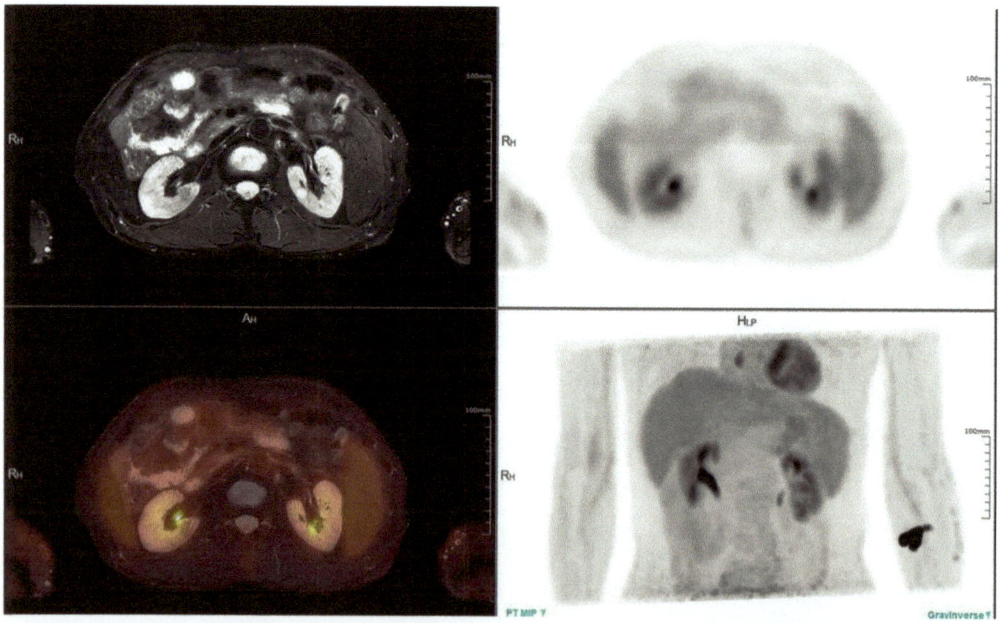

Figure 3.106 Renal axial T2WI (left upper), PET image (right upper), fusion image (left bottom), and MIP image (right bottom). There is a round hyperintensity signal on T2WI with hypometabolism on PET located in the left kidney, indicating a left renal cyst. *MIP*, Maximum Intensity Projection; *PET*, Positron Emission Tomography; *T2WI*, T2-Weighted Imaging.

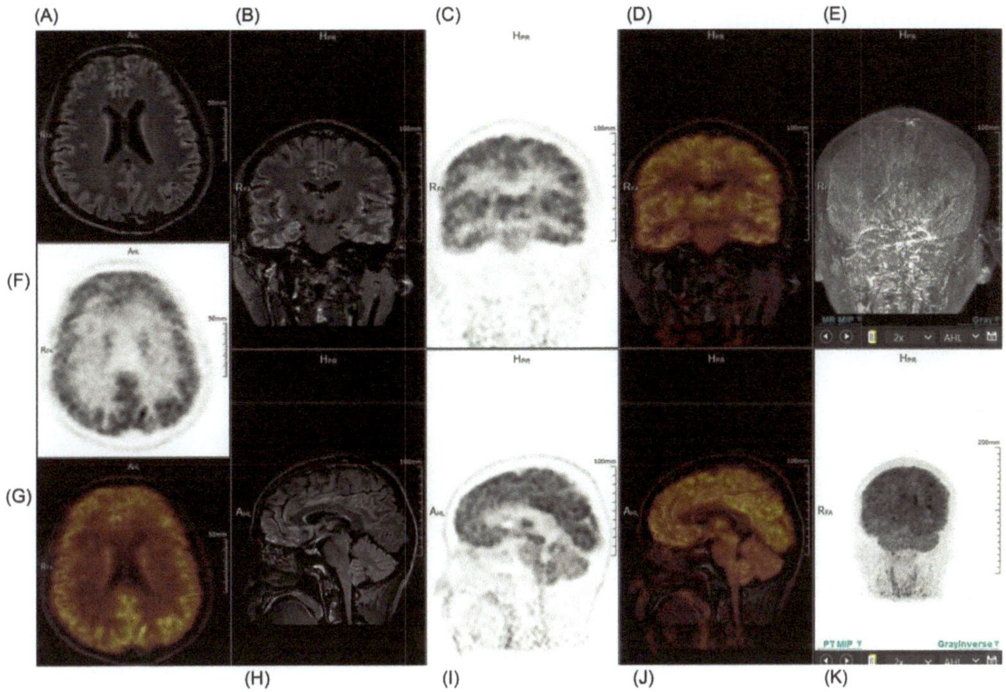

Figure 3.107 Brain 3D T2FLAIR (A, B, H), PET (C, F, I), fusion images (D, G, J) and MIP images (E, K). There are multiple patchy lesions with high signal intensity on T2FLAIR in the bilateral subcortical white matter area without abnormal metabolic activity, indicating ischemic demyelination.

Case 49

Medical history

A 48-year-old woman reported a fever of 38.5°C just 10 days prior to the admission, accompanied by cough with mild white sticky sputum. The patient went to the local hospital, and her blood tests showed decreased white blood cells and lymphocyte counts, increased myocardial enzymes, CRP, and ESR. The throat swab nucleic acid test was positive 6 days prior to the admission date.

Laboratory findings on admission

White blood cell count: 11.20×10^9/L, lymphocyte count: 1.75×10^9/L, lymphocyte percentage: 15.6%, C-reactive protein: 7.8 mg/L.

Laboratory findings 1 year and 7 months after discharge

White blood cell count: 7.65×10^9/L, lymphocyte count: 2.79×10^9/L, lymphocyte percentage: 36.5%, C-reactive protein: $<$0.499 mg/L.

Radiological findings

The radiological findings of chest CT scans during her hospitalization are as follows: Figs. 3.108–3.110.

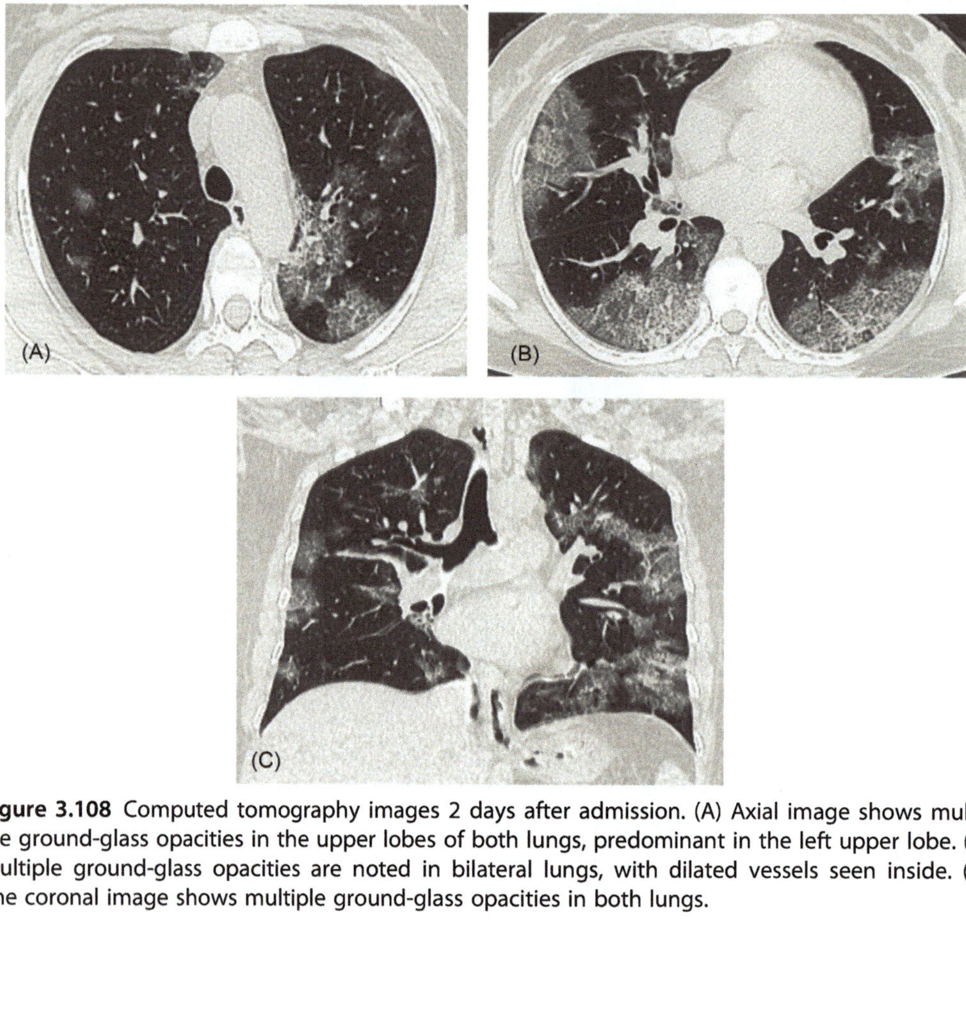

Figure 3.108 Computed tomography images 2 days after admission. (A) Axial image shows multiple ground-glass opacities in the upper lobes of both lungs, predominant in the left upper lobe. (B) Multiple ground-glass opacities are noted in bilateral lungs, with dilated vessels seen inside. (C) The coronal image shows multiple ground-glass opacities in both lungs.

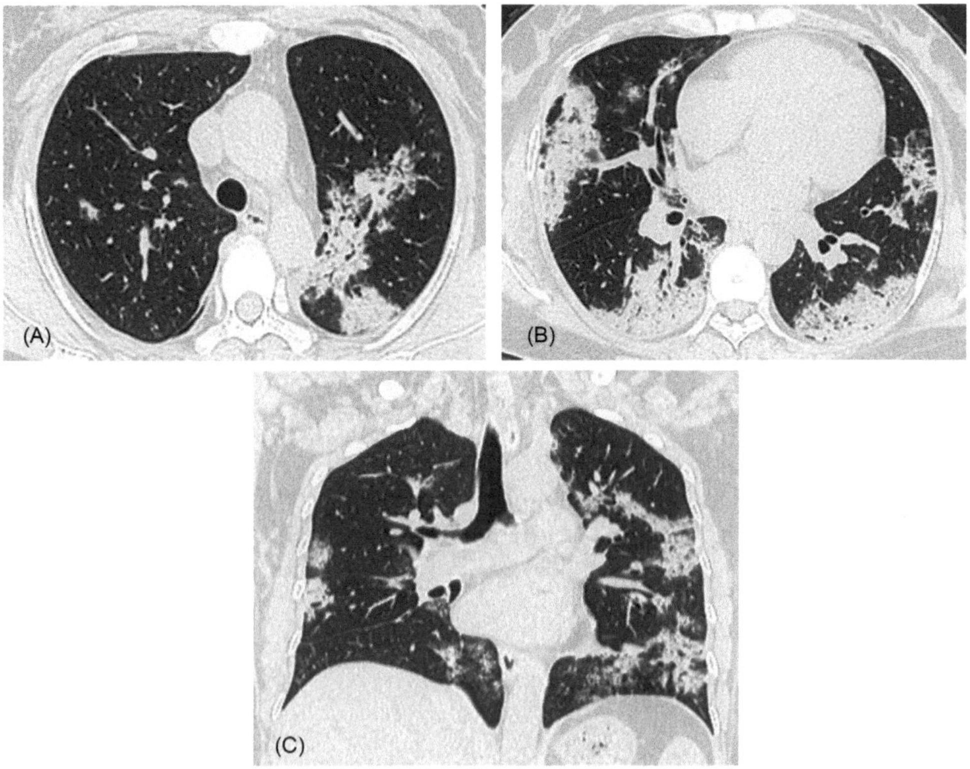

Figure 3.109 Computed tomography images 6 days after admission. Axial and coronal images show ground-glass opacities are partly absorbed, most of them having turned into consolidative opacities in both lungs. Bronchiectasis is seen in the lesions.

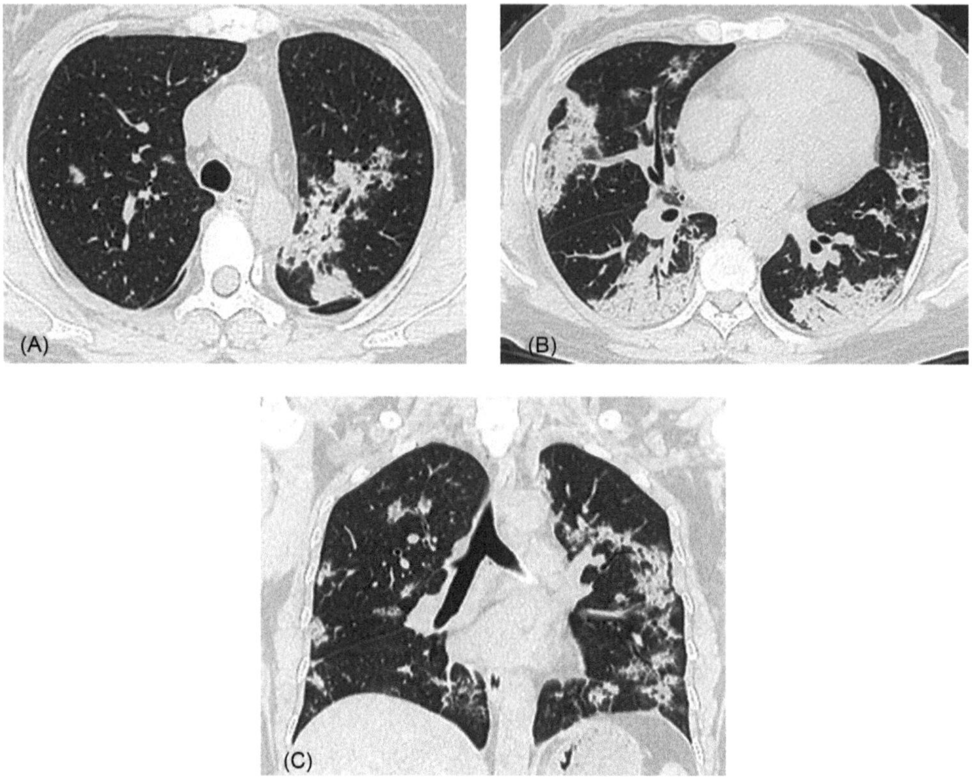

Figure 3.110 Computed tomography images 10 days after admission. Axial and coronal images show consolidative opacities in both lungs having been partially absorbed compared with prior images.

The radiological findings of chest CT scans after discharge are as follows: Figs. 3.111 and 3.112.

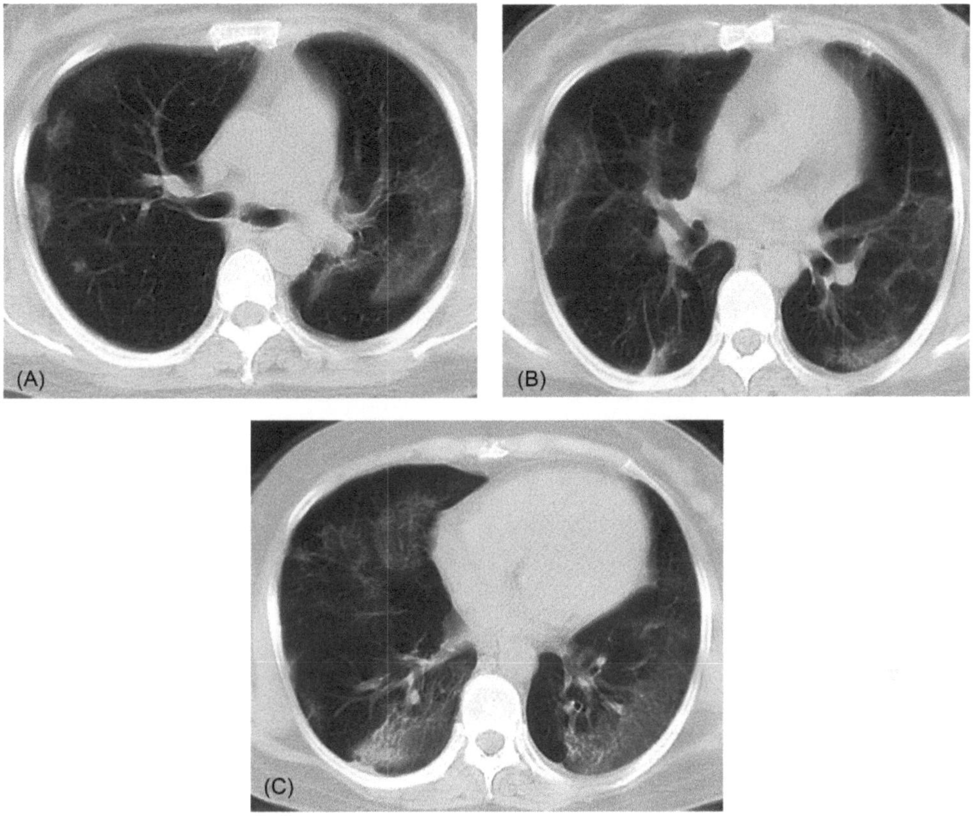

Figure 3.111 Chest computed tomography images 14 days after discharge. (A) Multiple ground-glass opacities are noted in the bilateral upper lobes. (B) Multiple ground-glass opacities and parenchymal bands are noted in the bilateral lungs. (C) Multiple ground-glass opacities are noted in the bilateral lungs.

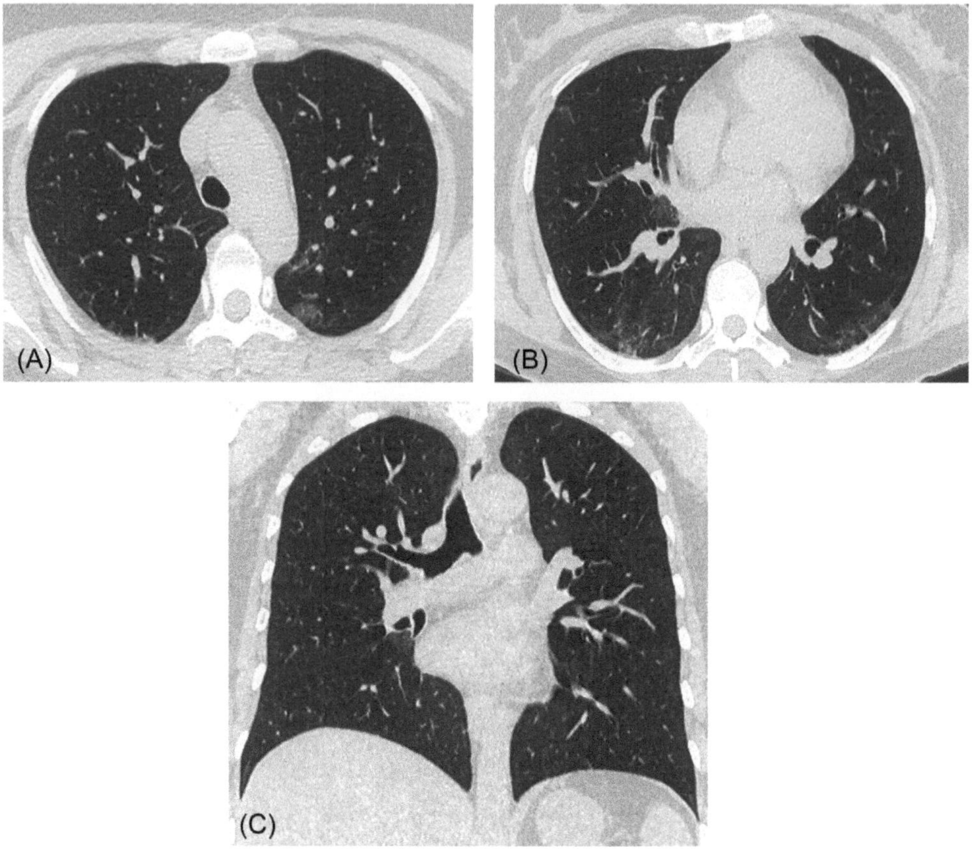

Figure 3.112 Chest computed tomography images 1 year and 7 months after discharge. Axial and coronal chest computed tomography images show small ground-glass opacities and a few fibrous bands remaining in bilateral lungs.

The patient underwent ^{18}F-FDG PET/CT, the image manifestations are as follows: Figs. 3.113–3.126.

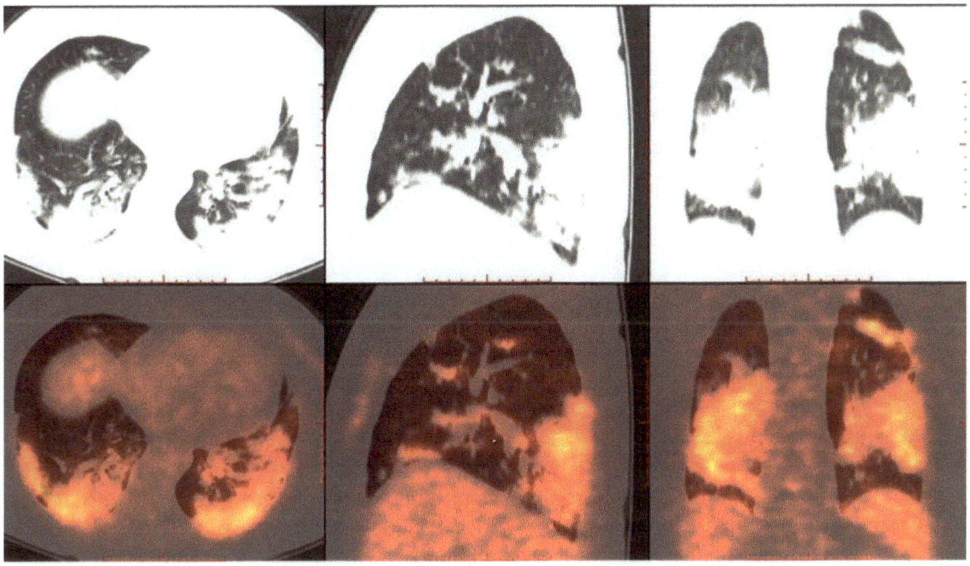

Figure 3.113 Computed tomography (top panel), PET/computed tomography fusion images (bottom panel) and 3D reconstruction images (lung window) 11 days after admission. There are multiple ground-glass and consolidative opacities in both lungs with increased metabolic activity. *PET*, Positron Emission Tomography.

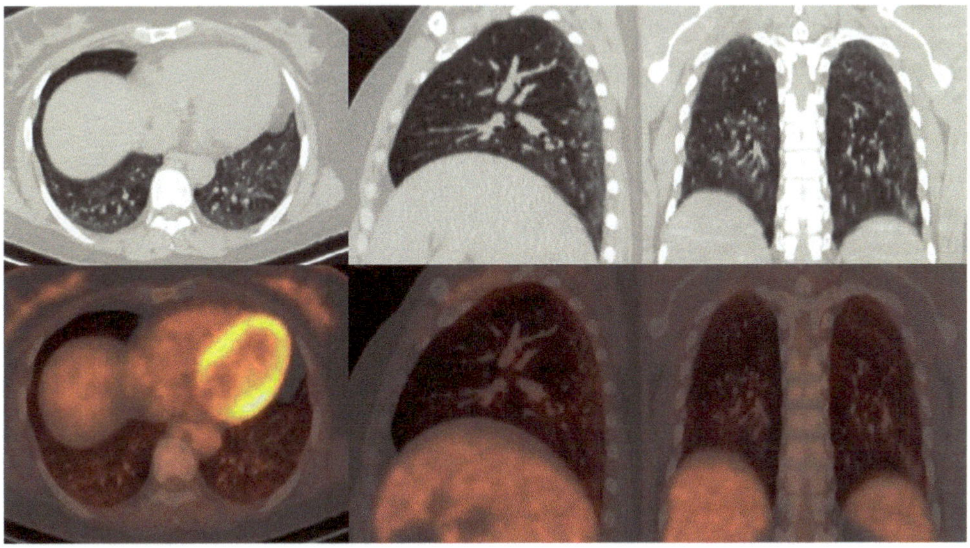

Figure 3.114 Computed tomography (top panel), PET/computed tomography fusion images (bottom panel), and 3D reconstruction images (lung window) 1 year and 7 months after discharge. The previous multiple ground-glass and consolidative opacities have disappeared in both lungs. There is no metabolic activity area in both lungs. *PET*, Positron Emission Tomography.

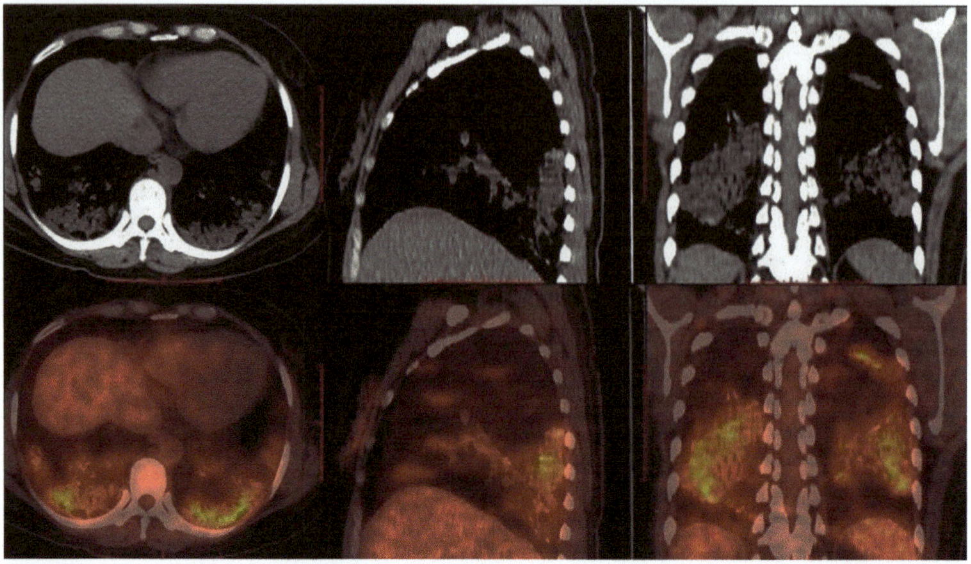

Figure 3.115 Computed tomography (top), PET/computed tomography fusion images (bottom) and 3D reconstruction images (mediastinum window) 11 days after admission. There are multiple ground-glass and consolidative opacities in both lungs with increased metabolic activity. *PET*, Positron Emission Tomography.

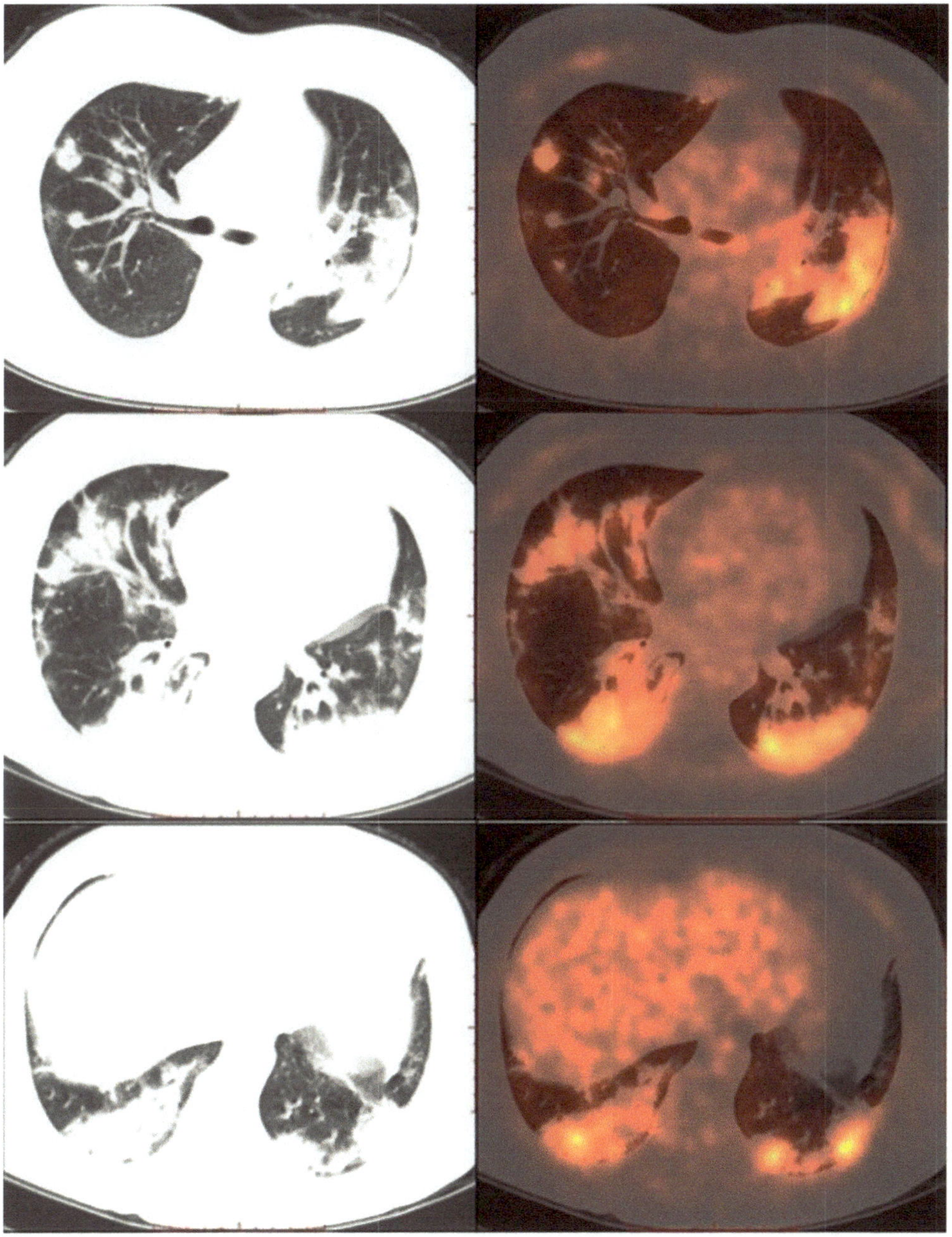

Figure 3.116 Computed tomography (left panel), PET/computed tomography fusion images (right panel) of multiple axial slices (lung window) 11 days after admission. There are multiple ground-glass and consolidative opacities in both lungs with increased metabolic activity. *PET*, Positron Emission Tomography.

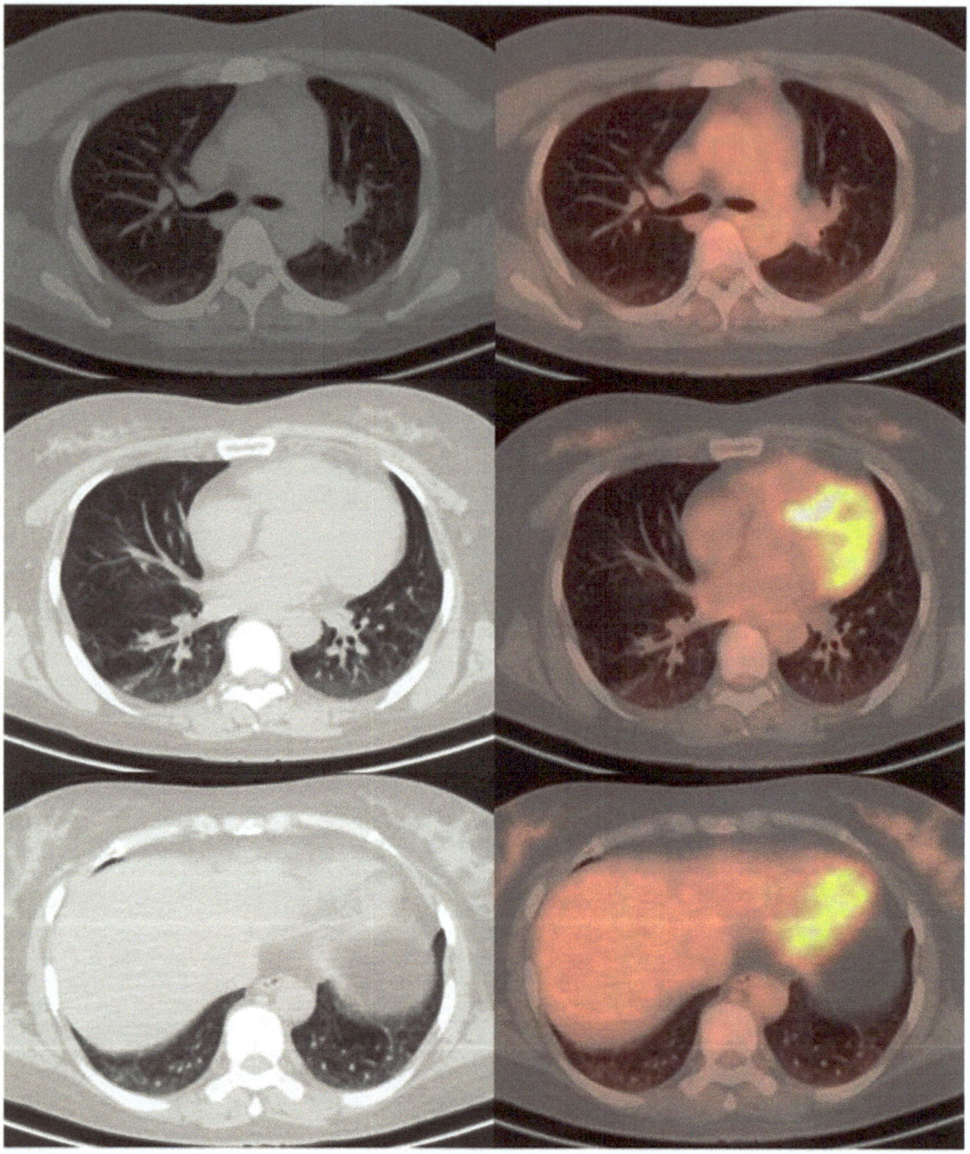

Figure 3.117 Computed tomography (left panel), PET/computed tomography fusion images (right panel) of multiple axial slices (lung window) 1 year and 7 months after discharge. The previous multiple ground-glass and consolidative opacities have disappeared in both lungs. There is no increased metabolic activity area in either lung. *PET*, Positron Emission Tomography.

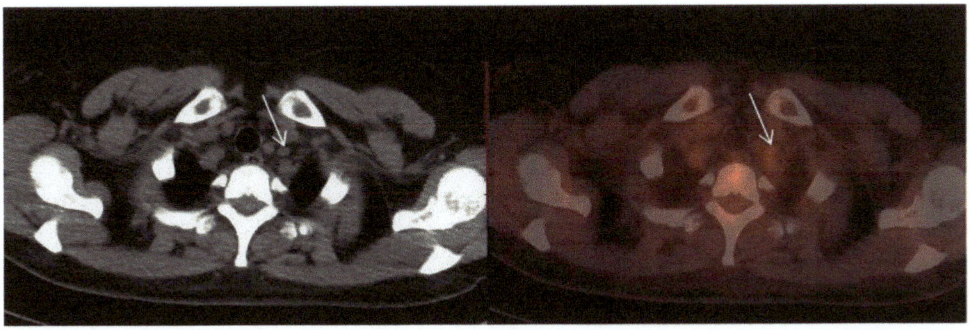

Figure 3.118 Computed tomography (left) and PET/computed tomography fusion images (right) 11 days after admission. There is a normal-sized lymph node with increased metabolic activity located in the left subclavian area (white arrows). *PET*, Positron Emission Tomography.

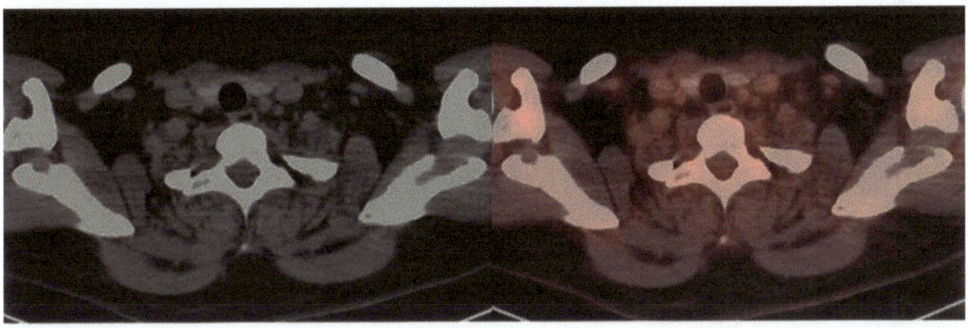

Figure 3.119 Computed tomography (left) and PET/computed tomography fusion images (right) 1 year and 7 months after discharge. There is a normal-sized lymph node in the left subclavian area with no increased metabolic activity. *PET*, Positron Emission Tomography.

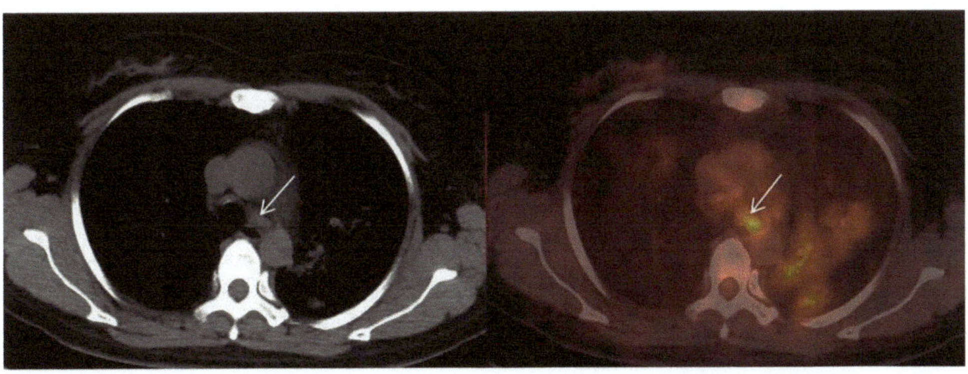

Figure 3.120 Computed tomography (left) and PET/computed tomography fusion images (right) 11 days after admission. There are multiple mild enlarged lymph nodes with increased metabolic activity located in the mediastinum (white arrows). *PET*, Positron Emission Tomography.

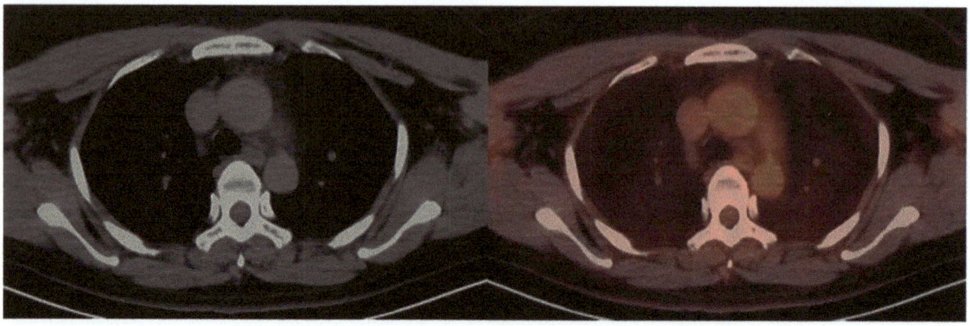

Figure 3.121 Computed tomography (left) and PET/computed tomography fusion images (right) 1 year and 7 months after discharge. There are normal-sized lymph nodes with no increased metabolic activity located in the mediastinum. *PET*, Positron Emission Tomography.

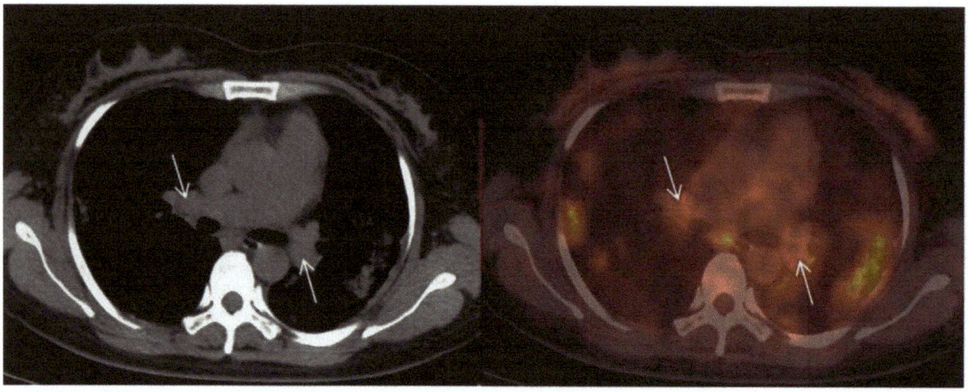

Figure 3.122 Computed tomography (left) and PET/computed tomography fusion images (right)11 days after admission. There are multiple mild enlarged lymph nodes with increased metabolic activity located in bilateral hilar regions (white arrows). *PET*, Positron Emission Tomography.

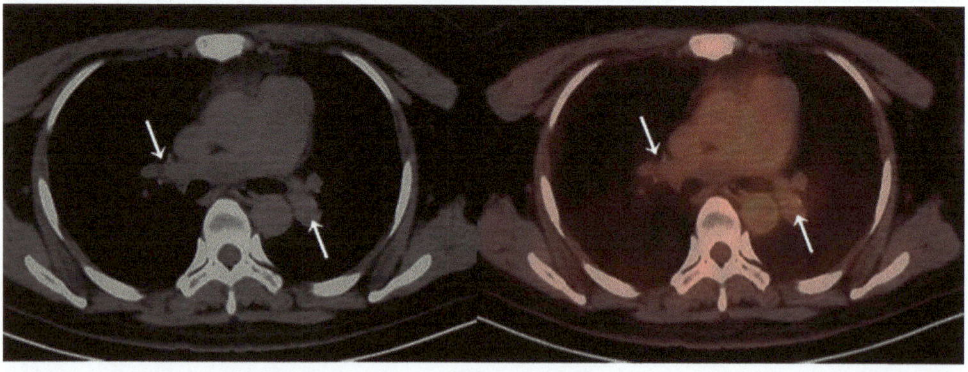

Figure 3.123 Computed tomography (left) and PET/computed tomography fusion images (right) 1 year and 7 months after discharge. There are normal-sized lymph nodes with no increased metabolic activity located in bilateral hilar regions (white arrows). *PET*, Positron Emission Tomography.

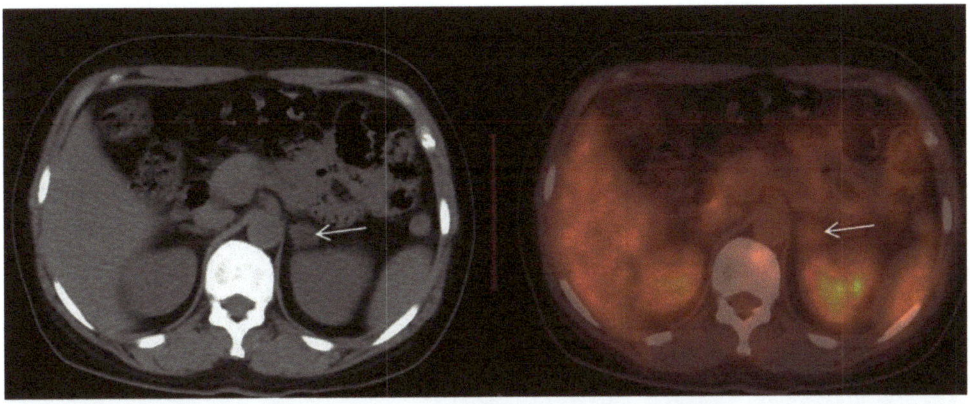

Figure 3.124 Computed tomography (left) and PET/computed tomography fusion images (right) 11 days after admission. There is a hypodense lesion in the left adrenal with normal metabolic activity, suggesting a benign lesion. *PET*, Positron Emission Tomography.

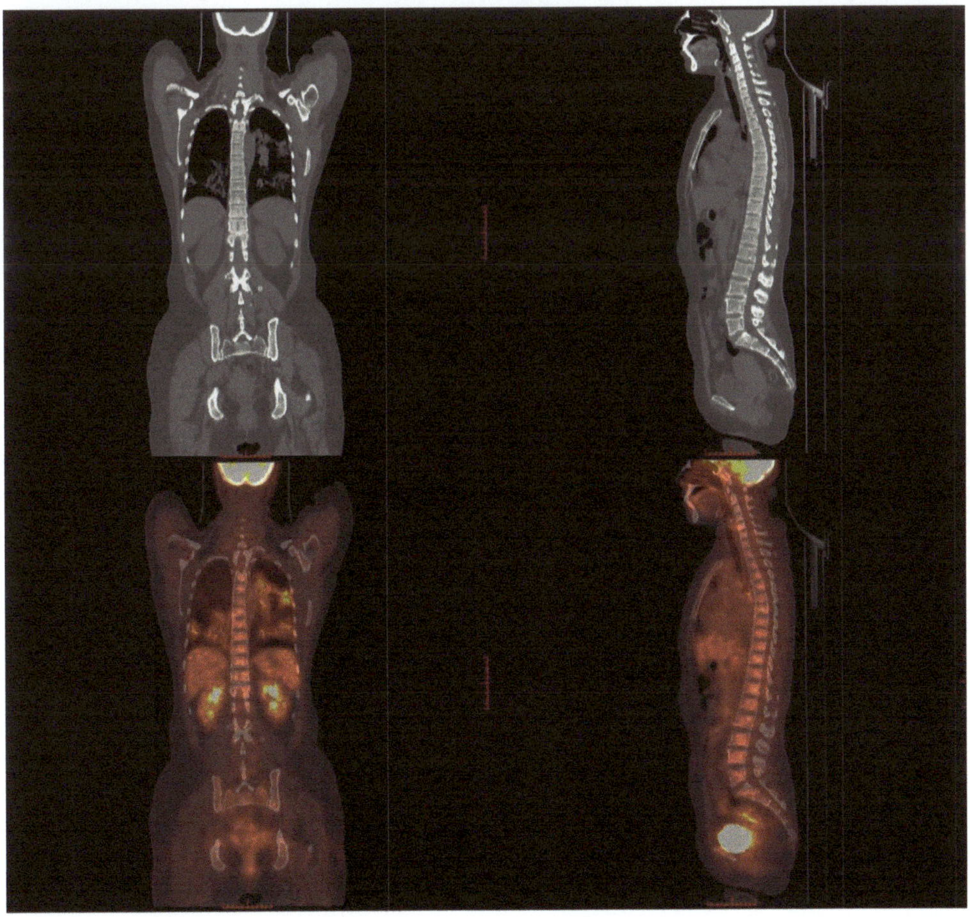

Figure 3.125 Computed tomography (left) and PET/computed tomography fusion images (right) 11 days after admission. There is diffusely increased metabolic activity in the vertebral marrow. *PET*, Positron Emission Tomography.

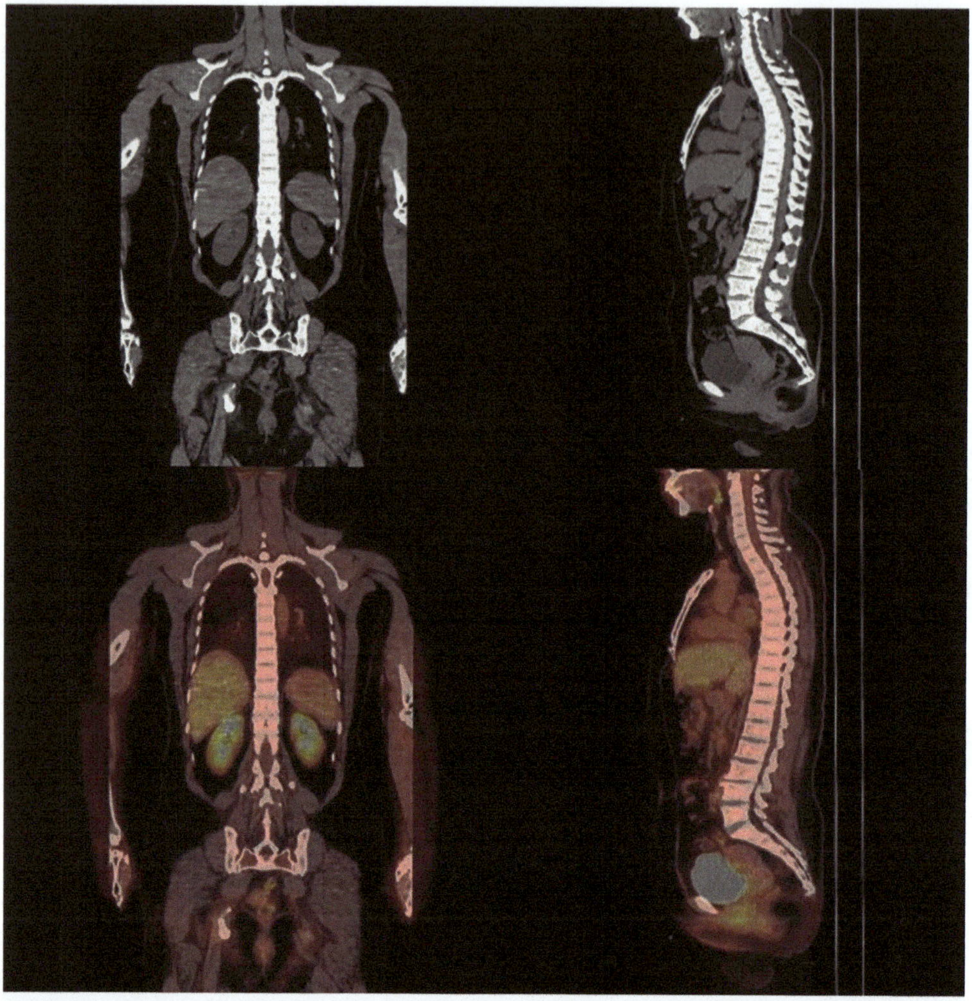

Figure 3.126 Computed tomography (left) and PET/computed tomography fusion images (right) 1 year and 7 months after discharge. There is no diffusely increased metabolic activity in the vertebral marrow. *PET*, Positron Emission Tomography.

Case 50

Medical history

A 59-year-old man reported fever 2 weeks prior to the admission, with a maximum temperature of 40°C, accompanied by a dry cough with little sputum. Infusion treatment was given to him at the local clinic for 7 days. The patient still had recurrent fever with a maximum temperature of 38°C, and he had a paroxysmal cough, expectoration with white sticky sputum, and chest tightness while walking. The nucleic acid test was positive 8 days prior to the admission date.

Laboratory findings on admission

White blood cell count: 7.00×10^9/L, lymphocyte count: 0.81×10^9/L, lymphocyte percentage: 11.6%, red blood cell count: 3.07×10^{12}/L, hemoglobin: 102.0 g/L, C-reactive protein: 2.3 mg/L.

Laboratory findings 1 year and 8 months after discharge

White blood cell count: 6.33×10^9/L, lymphocyte count: 2.69×10^9/L, lymphocyte percentage: 42.5%, red blood cell count: 4.42×10^{12}/L, hemoglobin: 141.0 g/L.

Radiological findings

The radiological findings of chest CT scans during his hospitalization are as follows: Figs. 3.127–3.130.

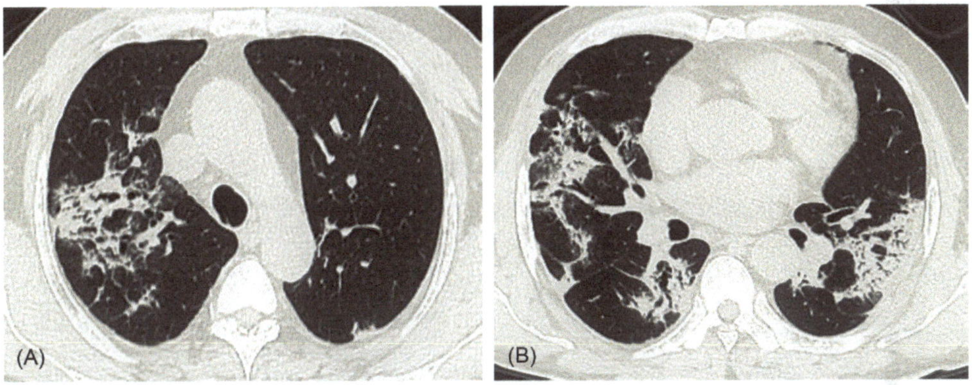

Figure 3.127 Computed tomography images 2 days after admission. (A) There are multiple consolidations in the right upper lobe, with parenchymal bands and bronchiectasis seen in the lesions. (B) There are multiple consolidations in both lungs, with parenchymal bands and bronchiectasis seen in the lesions.

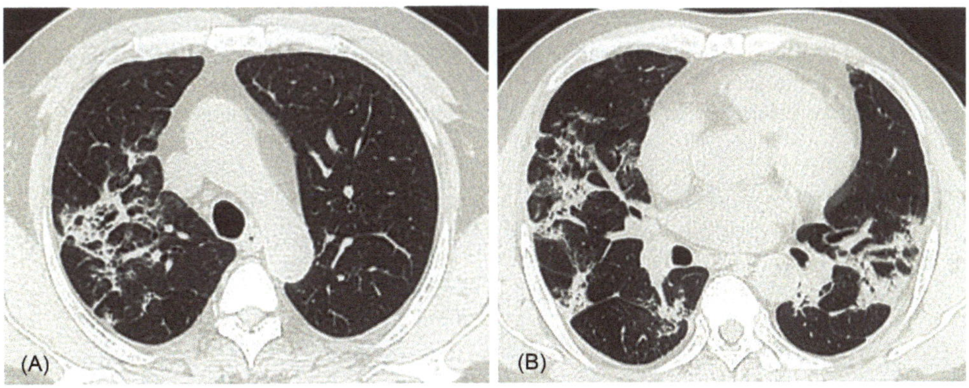

Figure 3.128 Computed tomography images 8 days after admission. (A) Mild absorption for the consolidations in the right upper lobe is noted. (B) The lesions in both lungs show slight absorption, with mild reduction in the area of the lesions.

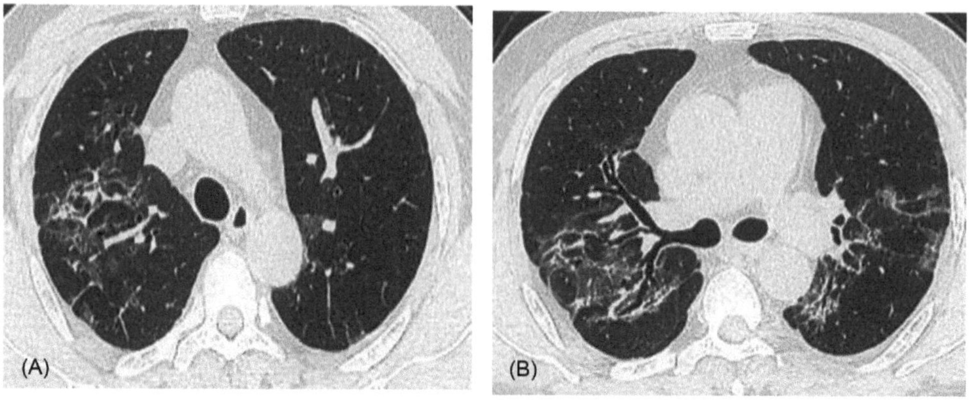

Figure 3.129 Chest computed tomography images 2 months after discharge. (A) There is absorption of the lesions in bilateral lungs, with remaining ground-glass opacities, parenchymal bands, and bronchiectasis. (B) There is absorption of the lesions in bilateral lungs, with remaining ground-glass opacities, parenchymal bands, and bronchiectasis.

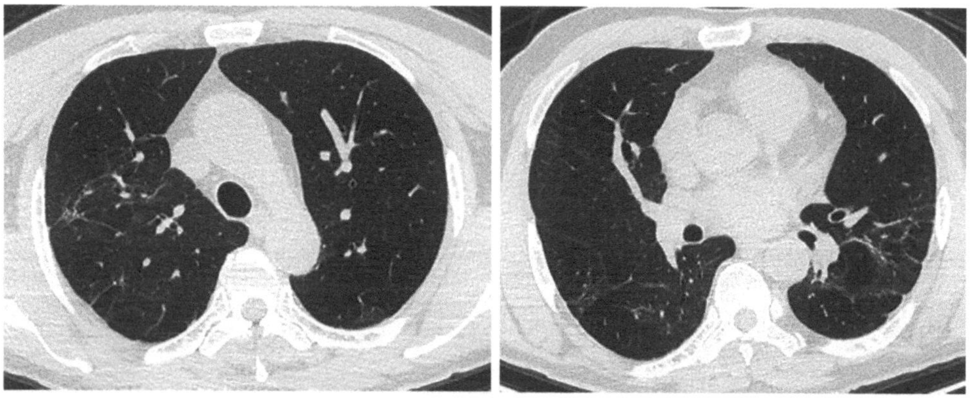

Figure 3.130 Chest computed tomography images 1 year and 8 months after discharge. (A) There is absorption for the lesions in bilateral lungs, with remaining parenchymal bands. (B) There is absorption for the lesions in bilateral lungs, with remaining parenchymal bands.

The radiological findings of chest CT scans after discharge are as follows:

The patient underwent ^{18}F-FDG PET/CT, and the image manifestations are as follows: Figs. 3.131–3.138.

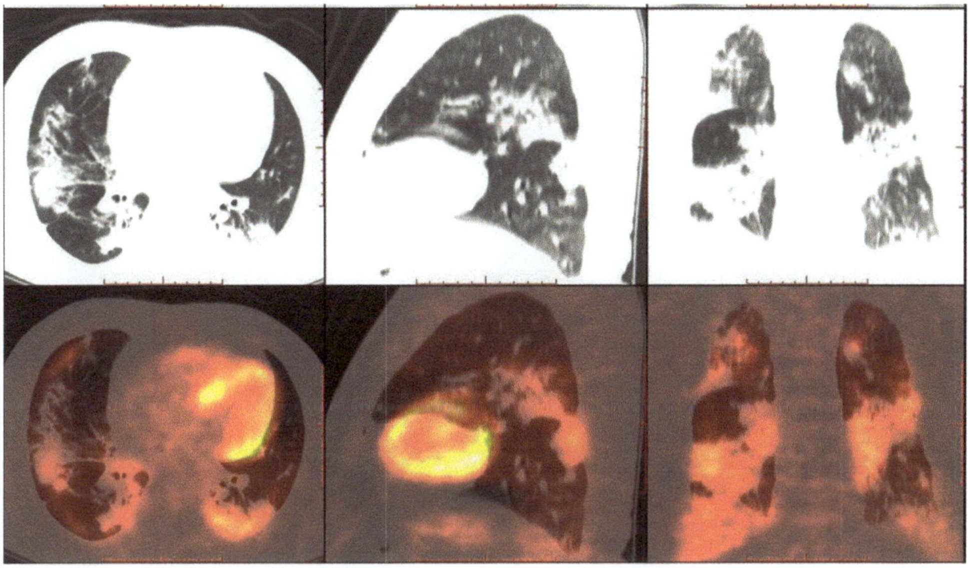

Figure 3.131 Computed tomography (top panel), PET/computed tomography fusion images (bottom panel), and 3D reconstruction images (lung window) 11 days after admission. There are multiple consolidative opacities and bands in both lungs with increased metabolic activity. *PET*, Positron Emission Tomography.

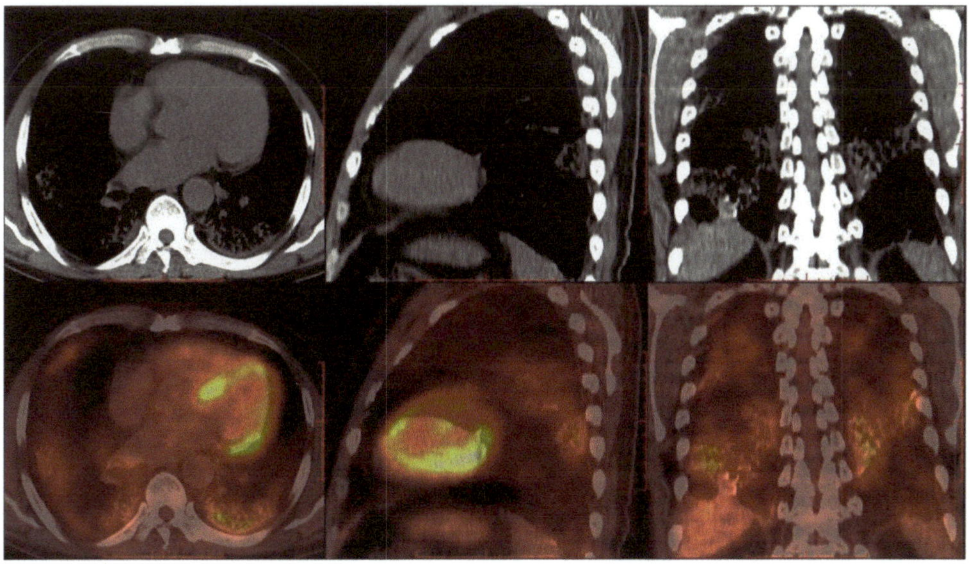

Figure 3.132 Computed tomography (top panel), PET/computed tomography fusion images (bottom panel), and 3D reconstruction images (mediastinum window) 11 days after admission. There are multiple consolidative opacities and bands in both lungs with increased metabolic activity. *PET*, Positron Emission Tomography.

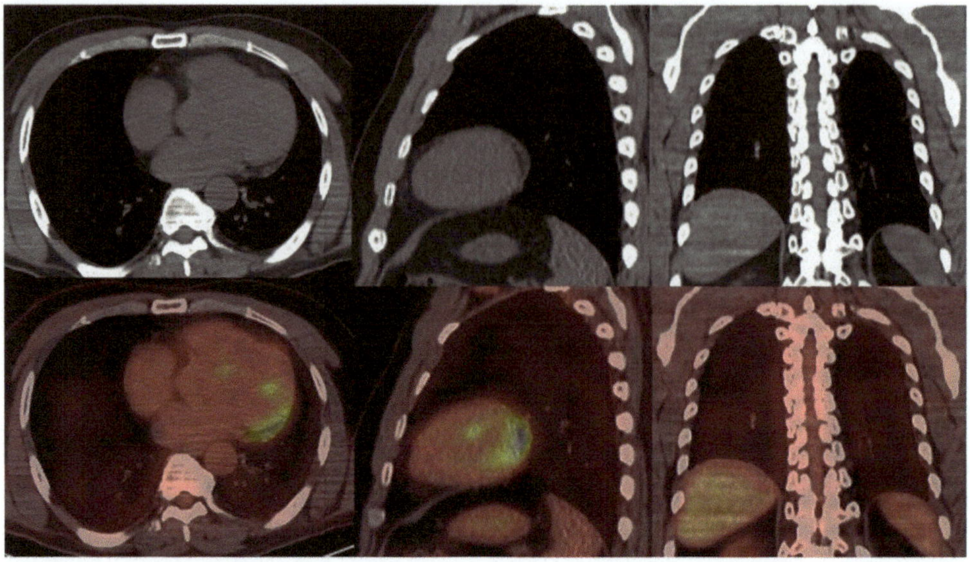

Figure 3.133 Computed tomography (top panel), PET/computed tomography fusion images (bottom panel), and 3D reconstruction images (mediastinum window) 1 year and 8 months after discharge. There is absorption for the lesions in both lungs. There is no increased metabolic activity area in both lungs. *PET*, Positron Emission Tomography.

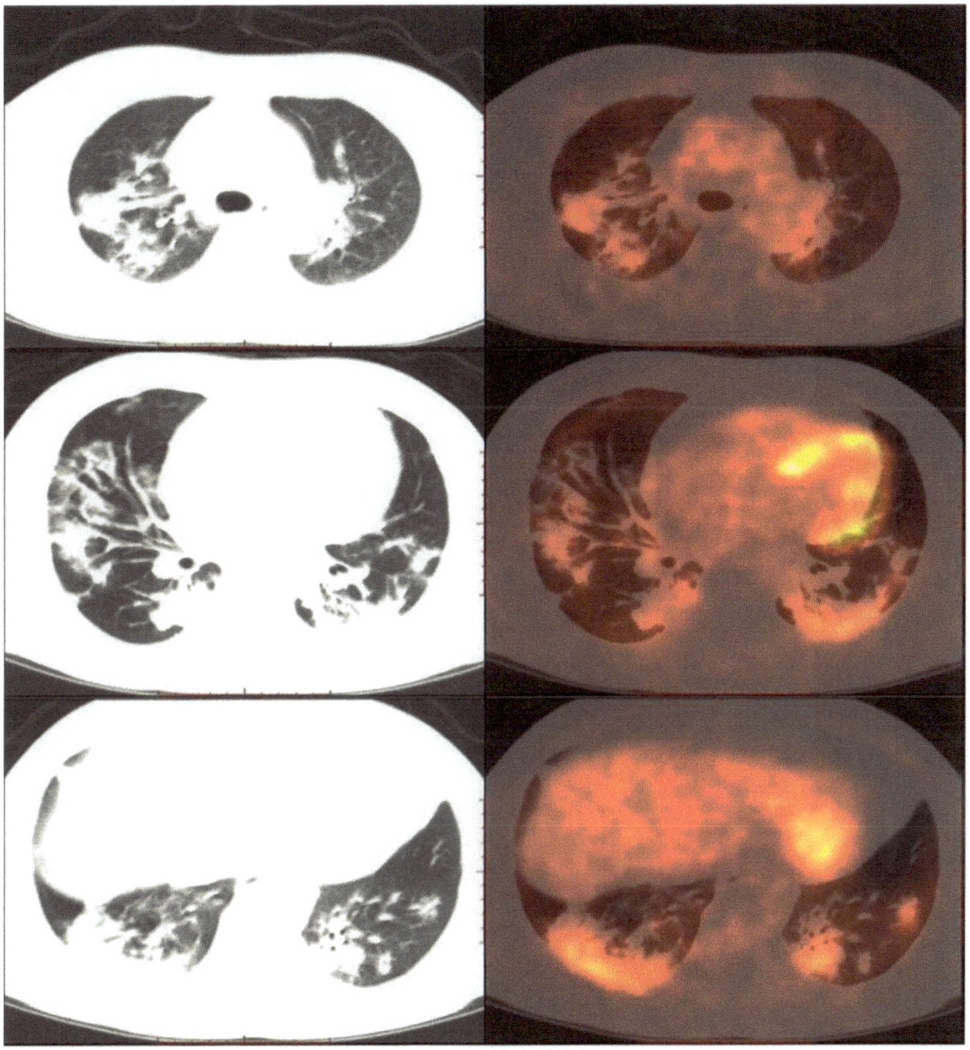

Figure 3.134 Computed tomography (left panel), PET/computed tomography fusion images (right panel) of multiple axial slices (lung window) 11 days after admission. There are multiple consolidative opacities and bands in both lungs with increased metabolic activity. *PET*, Positron Emission Tomography.

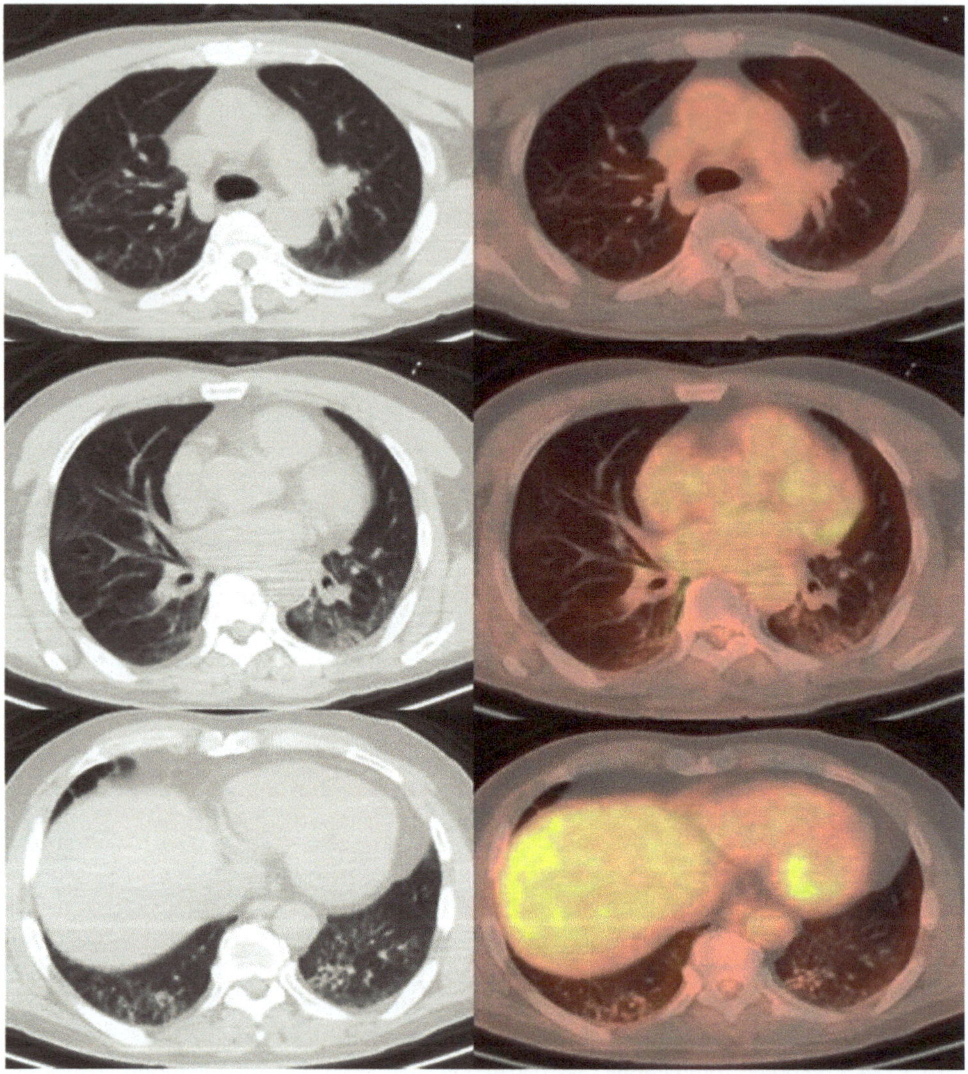

Figure 3.135 Computed tomography (left panel), PET/computed tomography fusion images (right panel) of multiple axial slices (lung window) 1 year and 8 months after discharge. There is absorption for the lesions in bilateral lungs, with remaining parenchymal bands. There is no increased metabolic activity area in either lung. *PET*, Positron Emission Tomography.

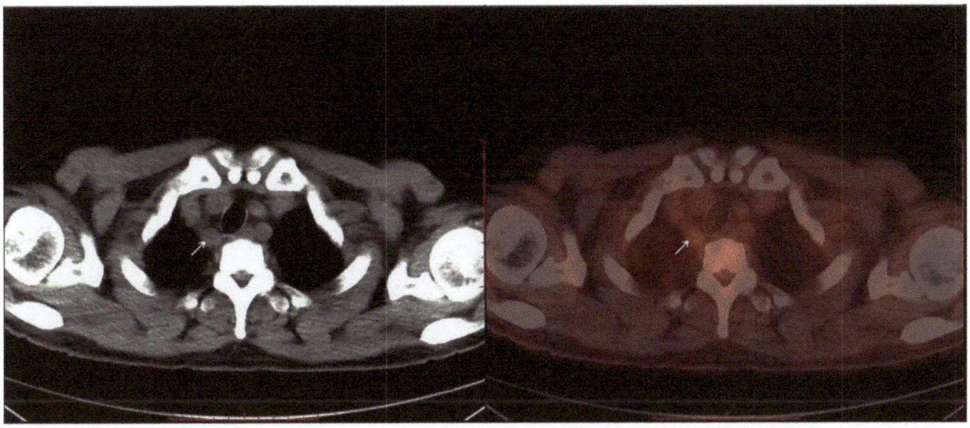

Figure 3.136 Computed tomography (left) and PET/computed tomography fusion images (right) 11 days after admission. There is a slightly enlarged lymph node on the right side of the upper mediastinal trachea with increased metabolic activity (white arrows). *PET*, Positron Emission Tomography.

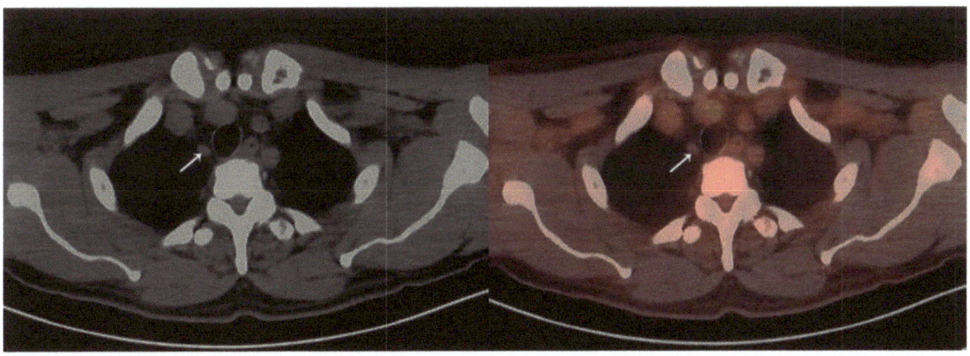

Figure 3.137 Computed tomography (left) and PET/computed tomography fusion images (right) 1 year and 8 months after discharge. There is a normal-sized lymph node on the right side of the upper mediastinal trachea with no increased metabolic activity (white arrows). *PET*, Positron Emission Tomography.

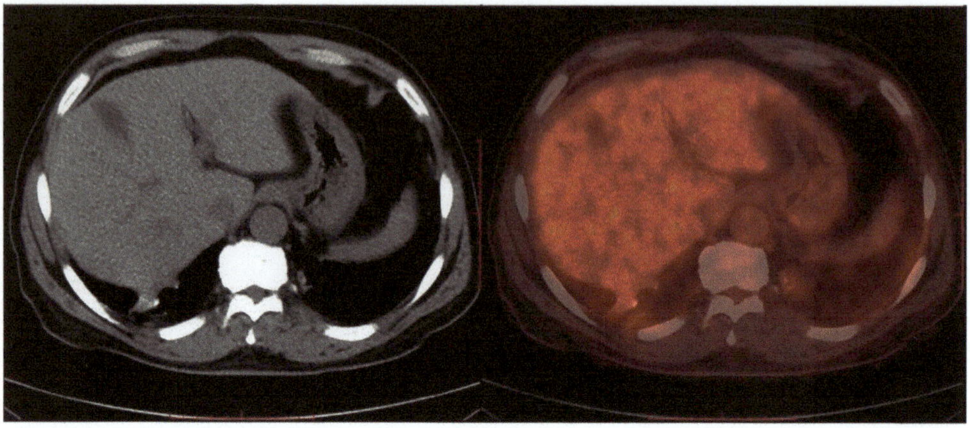

Figure 3.138 Computed tomography (left) and PET/computed tomography fusion images (right) 11 days after admission. There is no remarkable abnormal change in the liver and spleen. *PET*, Positron Emission Tomography.

CHAPTER 4

Bed-side DR and computed tomography manifestations and prognosis of critical COVID-19 cases

Case 51

Medical history

The patient is a 58-year-old man. He documented a history of contact with people diagnosed with COVID-19. He reported fever 11 days prior to the date of admission, with a maximum body temperature of 37.5°C, accompanied by a dry cough with white sticky sputum. He was diagnosed in the local clinic as having an upper respiratory infection with symptomatic treatment, though the symptoms did not go away. The throat swab nucleic acid test was positive 7 days prior to the admission.

Laboratory findings

White blood cell count: 11.00×10^9/L, lymphocyte count: 0.49×10^9/L, lymphocyte percentage: 4.4%, C-reactive protein: 41.5 mg/L.

Radiological findings

The radiological findings of bed-side DR during his hospitalization are as follows: Figs. 4.1–4.6.

Image Atlas of COVID-19
DOI: https://doi.org/10.1016/B978-0-443-18493-2.00004-8

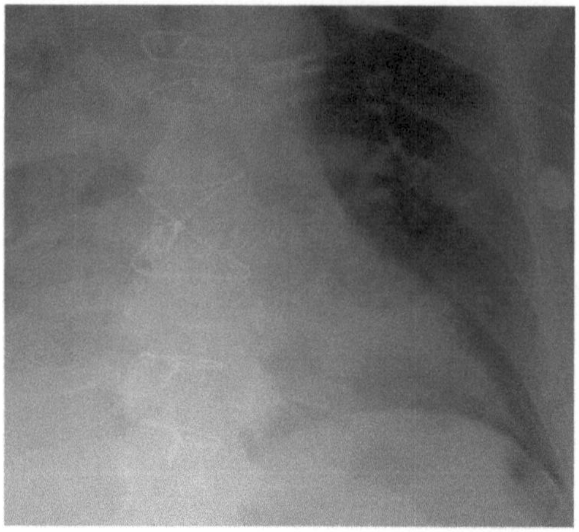

Figure 4.1 Bed-side DR on the day of admission. The right lung field show "white lung" change, the right costophrenic angle is blunt, and there are diffuse high-density shadows in the right lung.

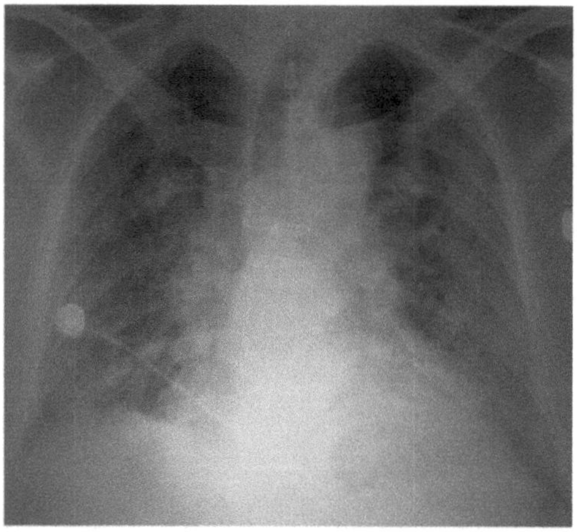

Figure 4.2 Bed-side DR 5 days after admission. The transparency of the right lung field is better compared with that of the prior image, whereas the lung markings of both lungs become obscure.

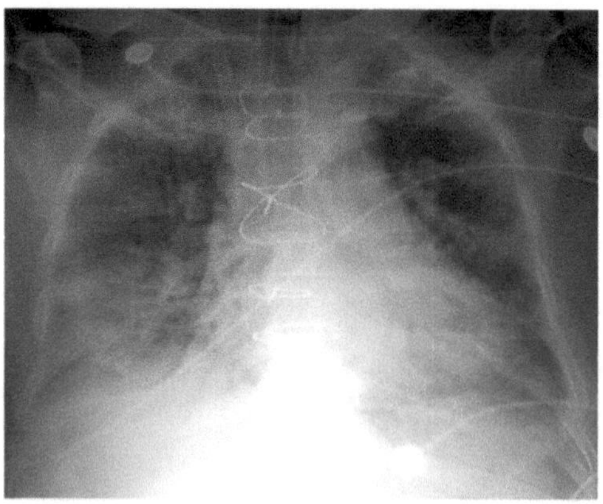

Figure 4.3 Bed-side DR 12 days after admission. The lesions in both lungs show progression, with new emerging patchy shadows in the right lung's lower field.

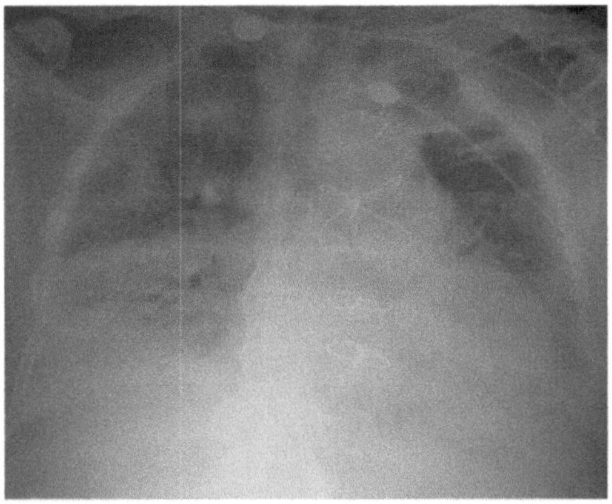

Figure 4.4 Bed-side DR 26 days after admission. The lesions in both lungs show progression, with white lung changes of both lung fields.

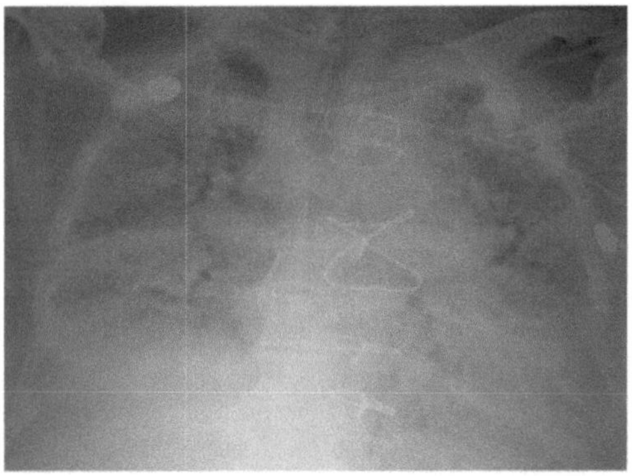

Figure 4.5 Bed-side DR 18 days after admission. The lesions in both lungs show progression, with the disappearance of bilateral costophrenic angle.

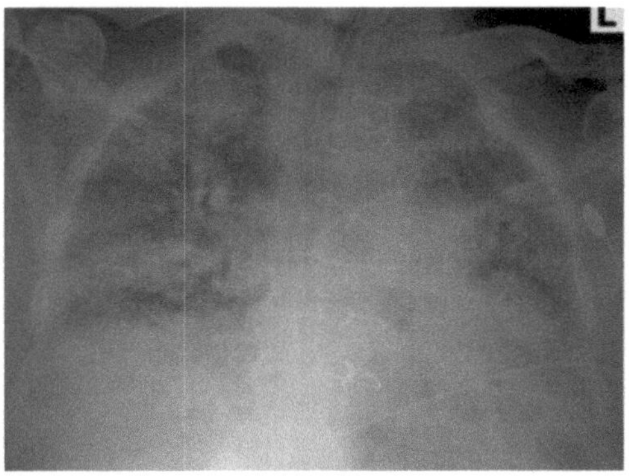

Figure 4.6 Bed-side DR 34 days after admission. The lesions in both lungs are similar to those of prior images.

Case 52

Medical history

The patient is an 87-year-old woman who documented a history of contact with people diagnosed with COVID-19. She reported fever 9 days prior to the date of admission, with a maximum body temperature of 37.8°C, accompanied by body ache in the whole body. The symptoms worsened 5 days prior to the admission, with a maximum body temperature of 38.8°C, accompanied by chest tightness and shortness of breath.

Laboratory findings

White blood cell count: 11.20×10^9/L, lymphocyte count: 0.57×10^9/L, lymphocyte percentage: 5.1%, C-reactive protein: 140.4 mg/L.

Radiological findings

The radiological findings of bed-side DR and chest computed tomography (CT) scans during her hospitalization are as follows: Figs. 4.7–4.11.

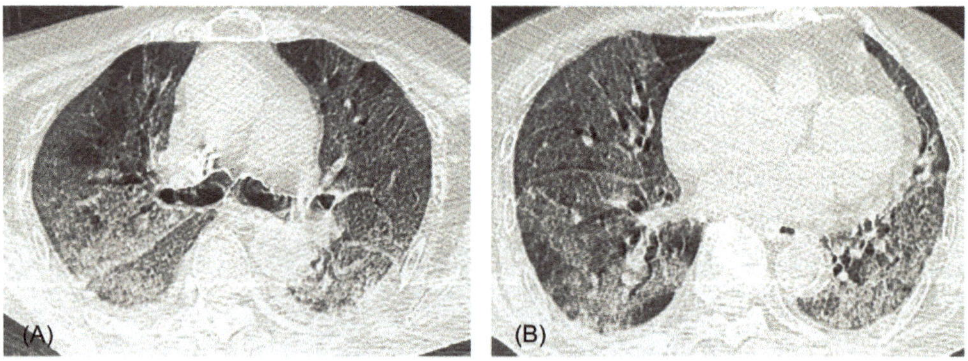

Figure 4.7 Chest computed tomography images on the day of admission. (A) Diffuse multiple ground-glass opacities in bilateral lungs and a small amount of bilateral pleural effusions are noted. (B) Diffuse multiple ground-glass opacities in the middle and lower lobes of the right lung and left lung, and a small amount of bilateral pleural effusions are noted.

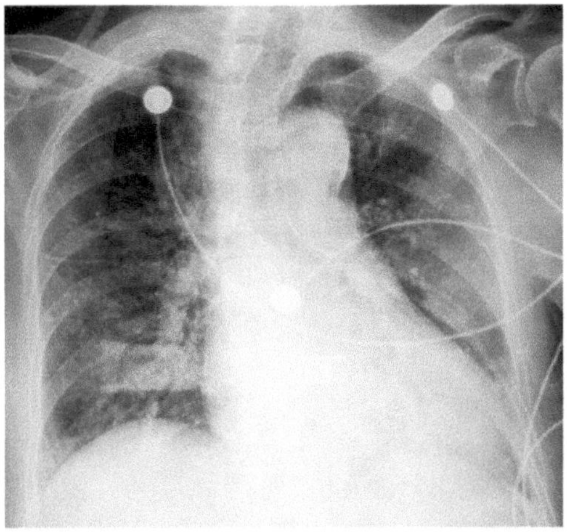

Figure 4.8 Bed-side DR on the day of admission. The lung markings of both lungs are obscure.

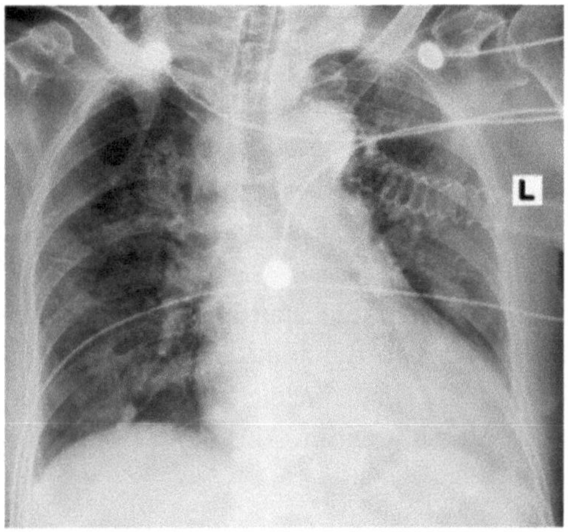

Figure 4.9 Bed-side DR 9 days after admission. The lesions in both lungs show progression, with a broader area of patchy shadows in both lung fields.

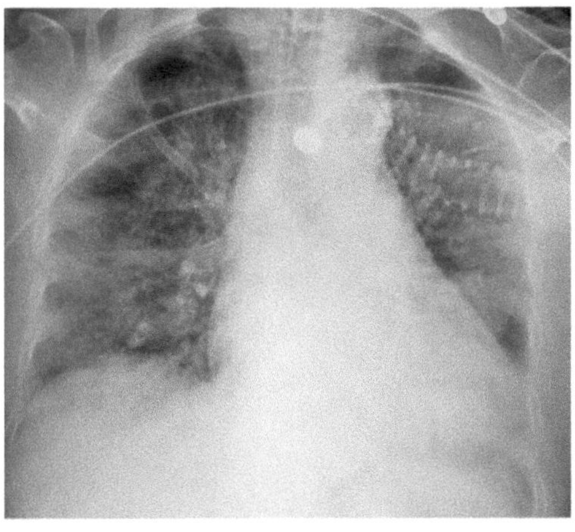

Figure 4.10 Bed-side DR 5 days after admission. The lesions in both lungs show progression, with obscurer lung markings and new emerging patchy shadows in both lung fields.

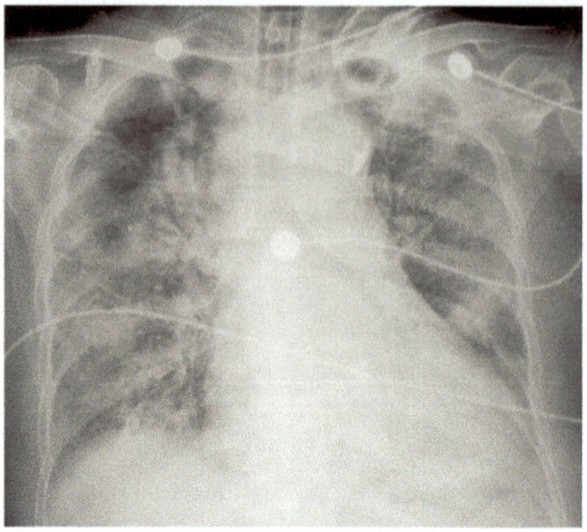

Figure 4.11 Bed-side DR 14 days after admission. The lesions in both lungs show remarkable progression, with an obvious decrease in the transparency of both lungs and enlarged patchy shadows in both lung fields.

Case 53

Medical history

The patient is an 86-year-old woman. She started to cough 8 days prior to the admission and reported fever just 5 days prior to the admission, with a maximum body temperature of 38.0°C, accompanied by chills. The nucleic acid test was positive 1 day prior to the admission date. She had intermittent delirium on the day of admission.

Laboratory findings

White blood cell count: $21.30 \times 10^9/L$, lymphocyte count: $1.06 \times 10^9/L$, lymphocyte percentage: 5.0%, C-reactive protein: 13.2 mg/L.

Radiological findings

The radiological findings of bed-side DR and chest CT scans during her hospitalization are as follows: Figs. 4.12 and 4.13.

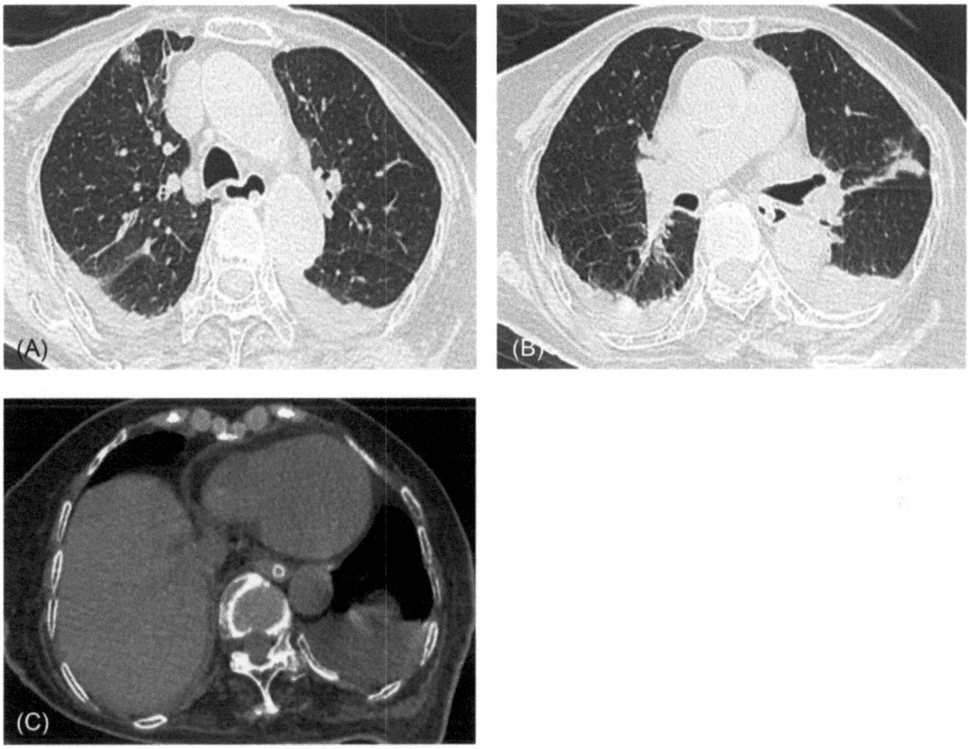

Figure 4.12 Chest computed tomography images on the day of admission. (A) There are ground-glass opacity and bands in the right lung's upper lobe. (B) There are consolidations in the left lung's upper lobe and bilateral pleural effusions. (C) Left pleural effusion is noted.

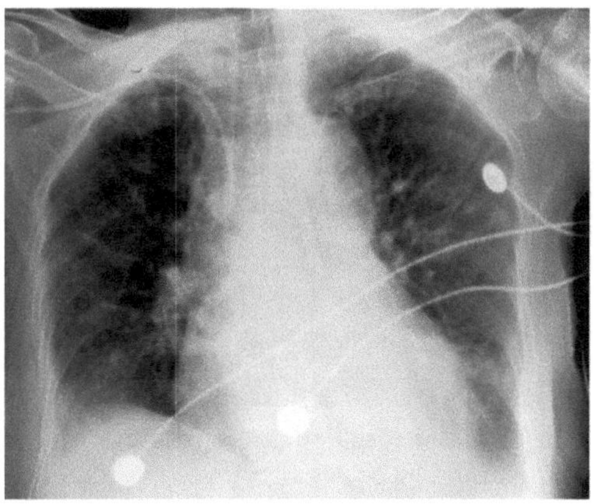

Figure 4.13 Bed-side DR 2 days after admission. The lung markings are obscure and there are patchy shadows in the outer zone of both lung fields. The bilateral costophrenic angle disappears.

Case 54

Medical history

The patient is a 74-year-old man who had been to the epidemic area. He reported fever 22 days prior to the date of admission, with a maximum temperature of 38.0°C. The nucleic acid test was positive 15 days prior to the admission date.

Laboratory findings

White blood cell count: $17.00 \times 10^9/L$, lymphocyte count: $0.51 \times 10^9/L$, lymphocyte percentage: 3.0%, C-reactive protein: 200 mg/L.

Radiological findings

The radiological findings of bed-side DR and chest CT scans during his hospitalization are as follows: Figs. 4.14–4.16.

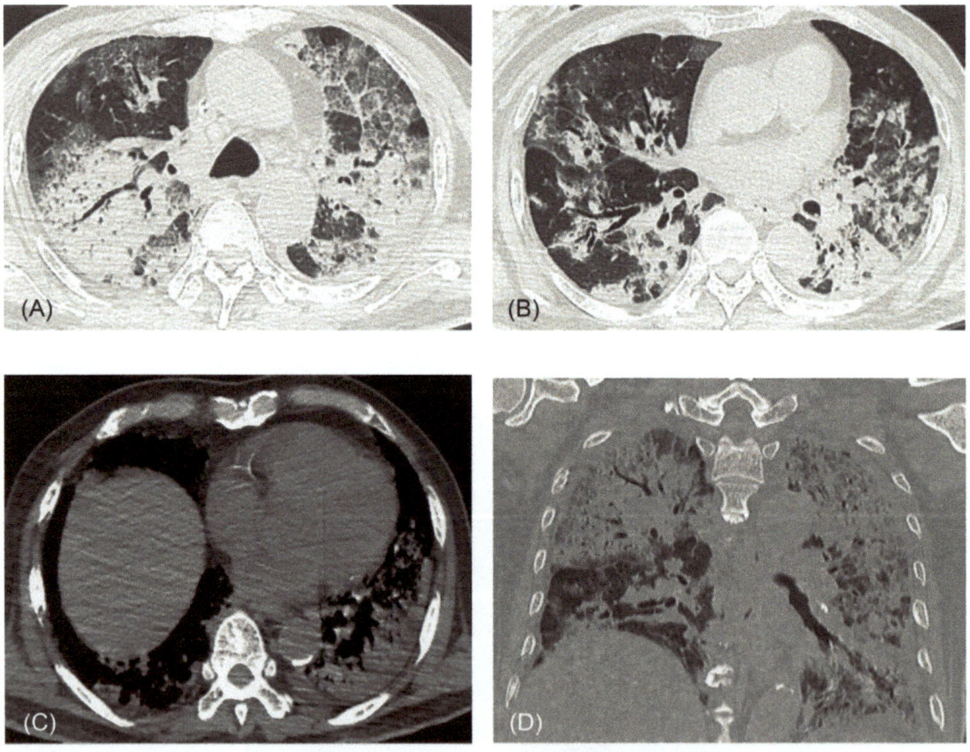

Figure 4.14 Computed tomography images on the day of admission. (A) Diffuse ground-glass opacities and consolidations are noted in both lungs, involving more than a 2/3 of the lung fields' area, with crazy-paving sign, air bronchogram, and distorted structure seen inside these lesions. (B) Diffuse ground-glass opacities and consolidations in both lungs, involving more than a 2/3 of the lung fields' area, with crazy-paving sign, an air bronchogram, and distorted structure seen inside these lesions. (C) Mild pleural effusion is noted in the left thoracic cavity. (D) Diffuse ground-glass opacities and consolidations in both lungs are noted.

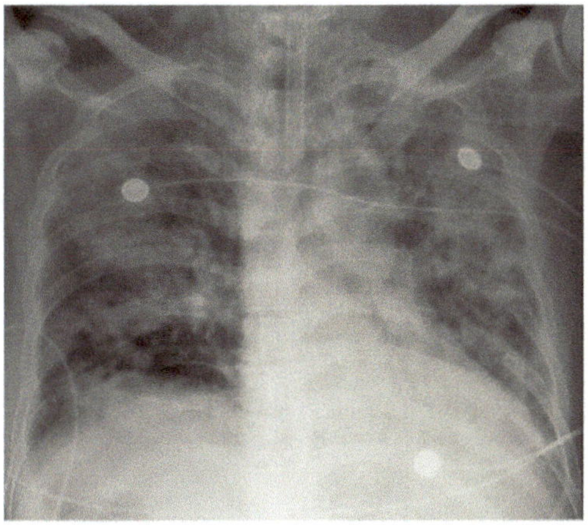

Figure 4.15 Bed-side DR on the day of admission. There are multiple patchy shadows in both lung fields.

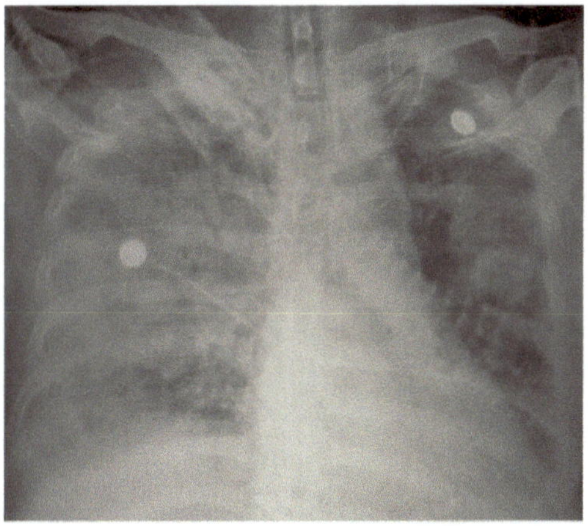

Figure 4.16 Bed-side DR 4 days after admission. The bilateral lung fields show "white lung" change, with much greater lesions compared with those of prior images.

Case 55

Medical history

The patient is a 57-year-old man who lives in the epidemic area. He reported fever 20 days prior to the admission date, with a maximum temperature of 37.4°C, accompanied by a cough with white sticky sputum. The nucleic acid test was positive 13 days prior to the admission. His blood oxygen saturation decreased 7 days prior to the admission. He had hemoptysis with about 150 mL of bright red blood. His blood oxygen saturation reduced to 85%, with a heart rate of 140–150 beats / min just 6 days prior to the admission date.

Laboratory findings

White blood cell count: $23.30 \times 10^9/L$, lymphocyte count: $1.16 \times 10^9/L$, lymphocyte percentage: 5.0%, C-reactive protein: 8.0 mg/L.

Radiological findings

The radiological findings of chest CT scans during his hospitalization are as follows: Figs. 4.17–4.19.

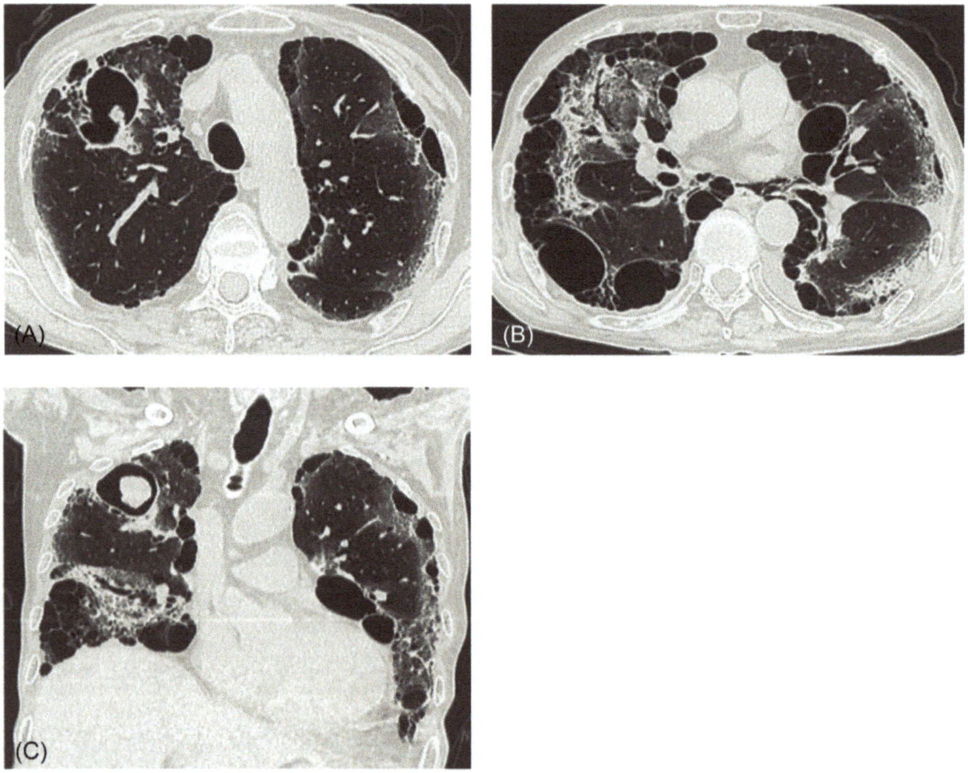

Figure 4.17 Chest computed tomography images 3 days after admission. (A) Bilateral lung emphysema with multiple subpleural pulmonary bullae are noted. (B) There are multiple groundglass opacities, consolidations and bands in both lungs, with crazy-paving sign and bronchiectasis seen inside. (C) A thin-walled cavity is seen in the anterior segment of the right lung's upper lobe, with an eccentric nodule in it indicating fungal infection.

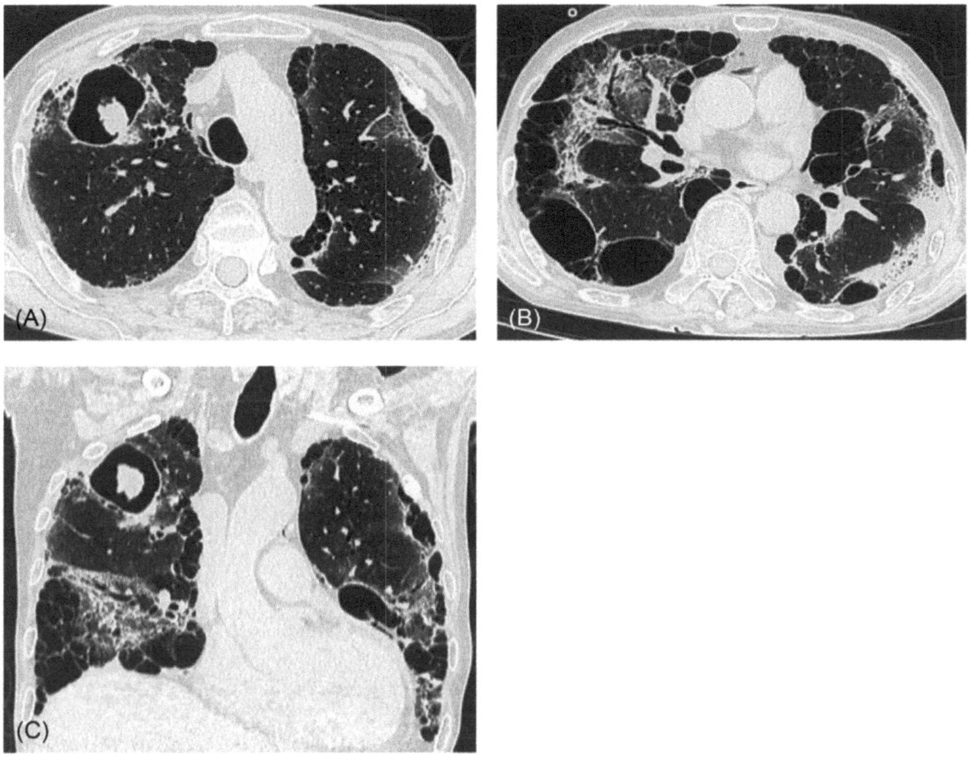

Figure 4.18 Chest computed tomography images 6 days after admission. (A) Bilateral lung emphysema with multiple subpleural pulmonary bullae are noted. (B) There are multiple groundglass opacities, consolidations and bands in both lungs, with crazy-paving sign and bronchiectasis seen inside. (C) A thin-walled cavity is noted in the anterior segment of the right lung's upper lobe, with an eccentric nodule in it indicating fungal infection. The lesions are similar with prior images.

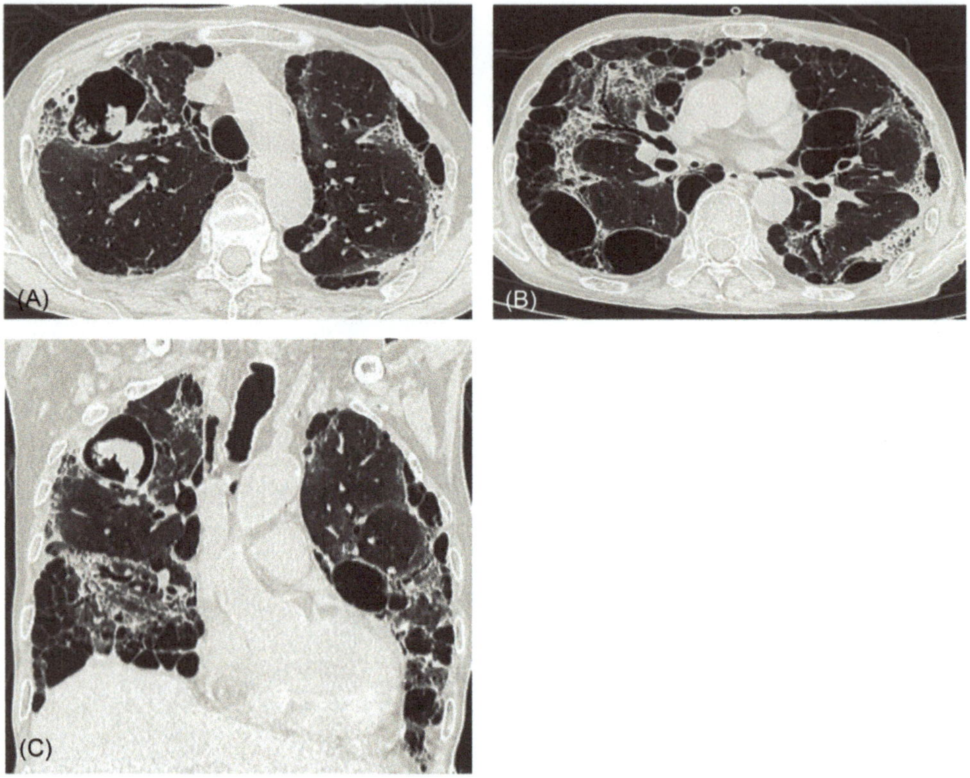

Figure 4.19 Chest computed tomography images 14 days after admission. (A) The lesions in bilateral lungs show slight absorption compared with that of prior images. (B) The lesions in bilateral lungs show slight absorption compared with that of prior images. (C) The lesions in bilateral lungs show slight absorption compared with that of prior images.

Case 56

Medical history

The patient is an 81-year-old man. He documented a history of contact with people diagnosed with COVID-19. He started to cough 17 days prior to the admission date. The nucleic acid test was positive 13 days prior to the admission date. He reported fever with a maximum temperature of 38.2°C, accompanied by worsened spates of cough and chest tightness.

Laboratory findings

White blood cell count: 4.00×10^9/L, lymphocyte count: 1.06×10^9/L, lymphocyte percentage: 26.3%, C-reactive protein: 46.4 mg/L.

Radiological findings

The radiological findings of bed-side DR and chest CT scans during his hospitalization are as follows: Figs. 4.20–4.25.

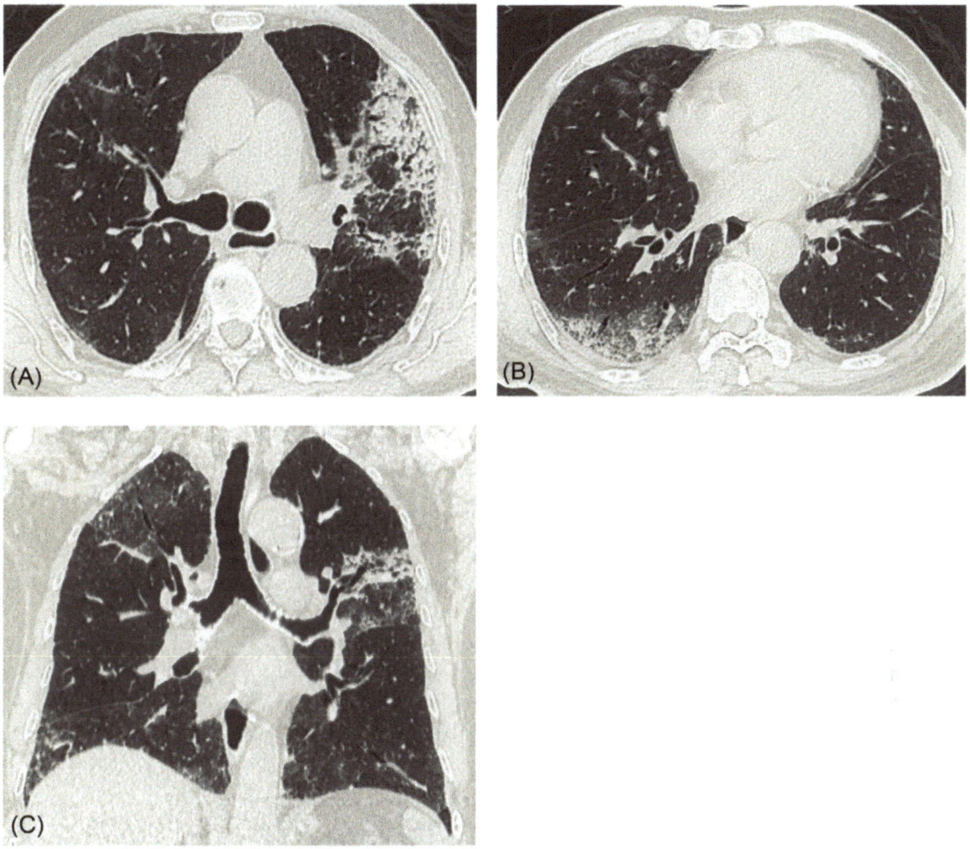

Figure 4.20 Chest computed tomography images 2 days after admission. (A) Multiple ground-glass opacities and consolidations are noted in bilateral lungs, with crazy-paving sign and bronchiectasis seen in the lesions, distributed predominantly in the peripheral area. (B) Multiple ground-glass opacities and consolidations are noted in bilateral lungs. (C) The coronal image shows multiple ground-glass opacities and consolidations in both lungs, with bronchiectasis seen inside.

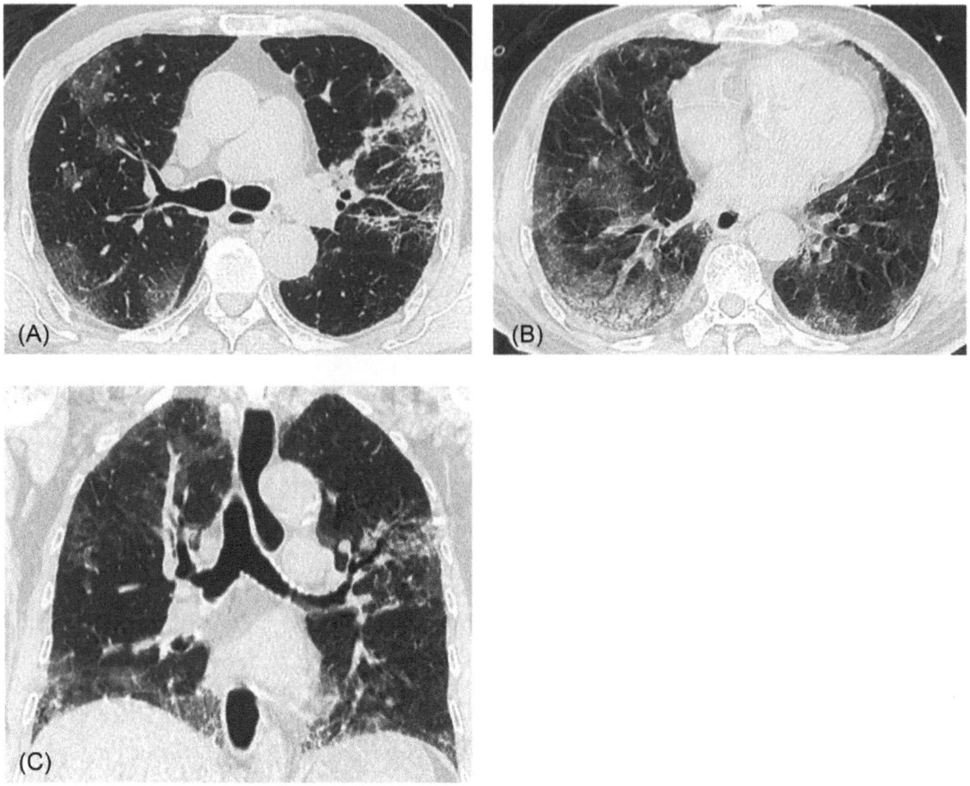

Figure 4.21 Chest computed tomography images 6 days after admission. (A) Compared with the previous images, consolidations in the left upper lung show mild absorption. (B) Absorption is noted in the right lower lobe, with reduction in attenuation. (C) The coronal image shows mild absorption in the left upper lobe and increase of ground-glass opacities in the bilateral lower lobes.

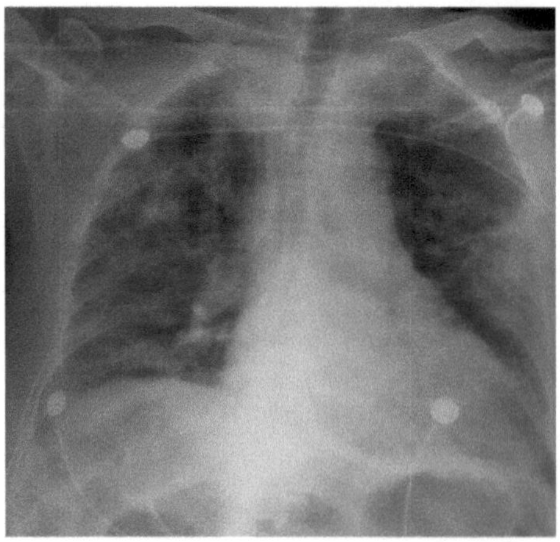

Figure 4.22 Bed-side DR 9 days after admission. There are multiple patchy shadows in both lung fields, with obscure lung markings.

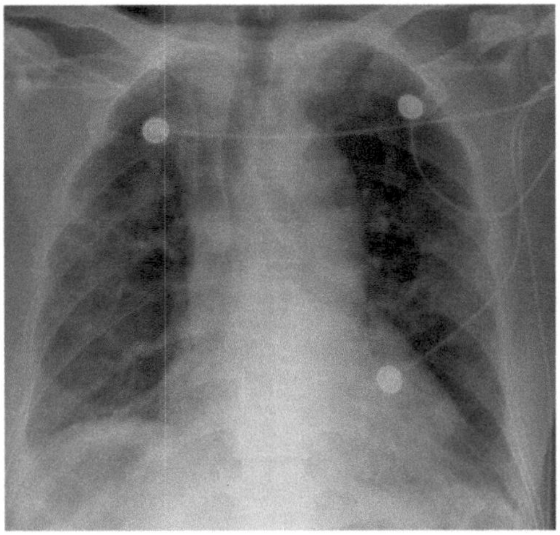

Figure 4.23 Bed-side DR 12 days after admission. There is slight absorption for the patchy shadows in both lung fields.

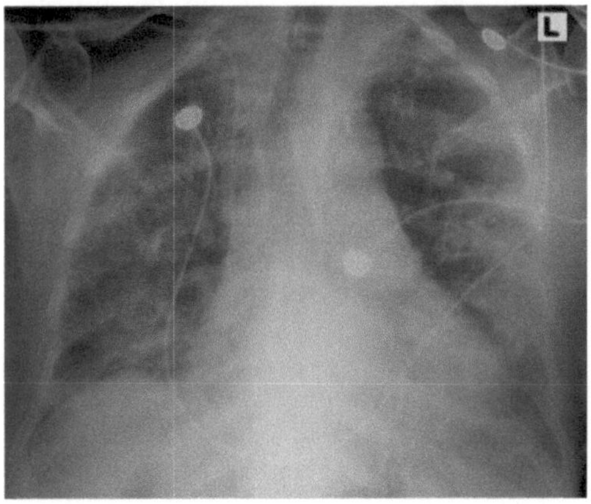

Figure 4.24 Bed-side DR 15 days after admission. The lesions show progression, with decreased transparency for the lung fields and the patchy shadows in both lung fields having enlarged.

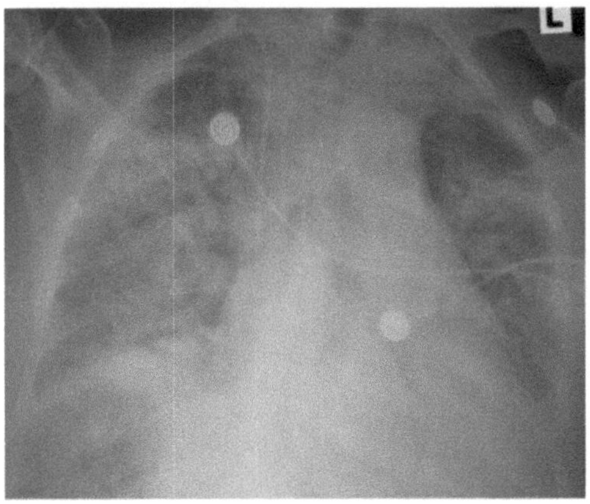

Figure 4.25 Bed-side DR 19 days after admission. The bilateral lung fields show "white lung" change, with much greater lesions compared with those of prior images.

Case 57

Medical history

The patient is an 89-year-old man who documented a history of contact with people diagnosed with COVID-19. He reported fever and fatigue just 4 days prior to the date of admission, with a maximum body temperature of 37.7°C.

Laboratory findings

White blood cell count: $3.20 \times 10^9/L$, lymphocyte count: $0.28 \times 10^9/L$, lymphocyte percentage: 8.7%, C-reactive protein: 126.3 mg/L.

Radiological findings

The radiological findings of bed-side DR during his hospitalization are as follows: Figs. 4.26–4.29.

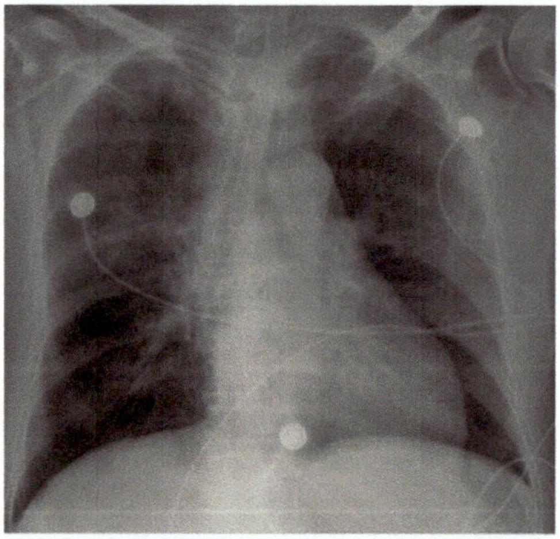

Figure 4.26 Bed-side DR 3 days after admission. There are multiple patchy shadows in the bilateral upper lung fields, with obscure lung markings.

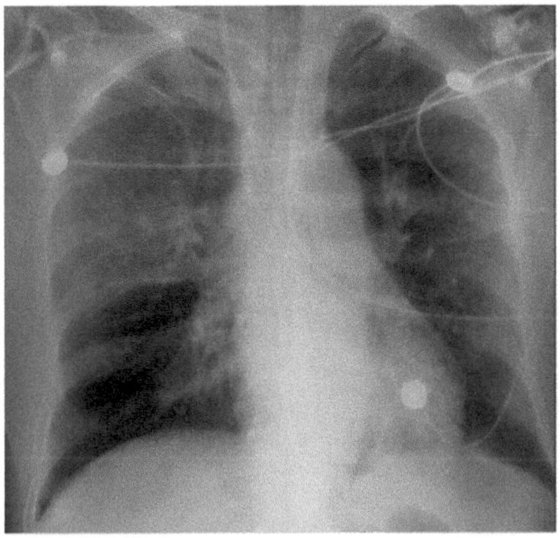

Figure 4.27 Bed-side DR 6 days after admission. The lesions show progression compared with those of the prior image, especially in the right lung's upper lobe.

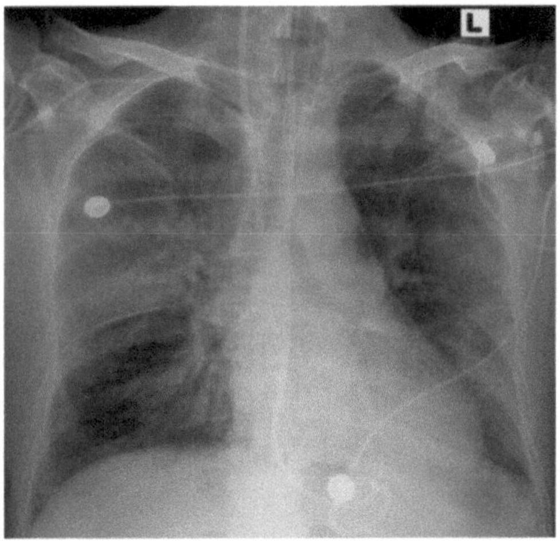

Figure 4.28 Bed-side DR 10 days after admission. The lesions show continuing progression compared with those of the prior images, with more patchy shadows in the bilateral lung fields.

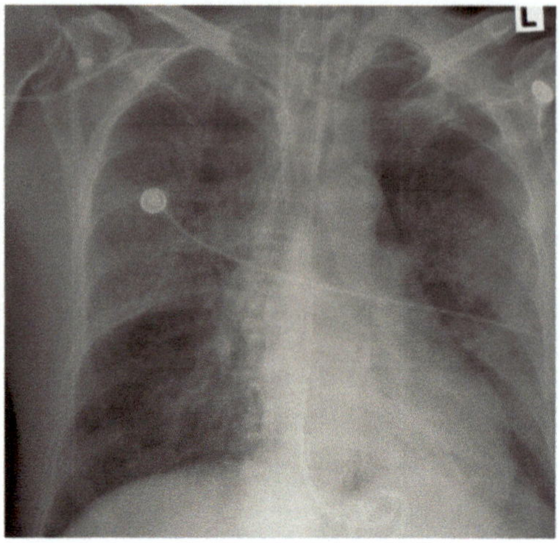

Figure 4.29 Bed-side DR 12 days after admission. The bilateral lung fields show "white lung" change, with much greater lesions compared with those of the prior image.

Case 58

Medical history

The patient is a 61-year-old woman who documented a history of travel and reported fever 21 days prior to the date of admission, with a maximum body temperature of 38.5°C, accompanied by cough, chills, muscle soreness, and fatigue. The nucleic acid test was positive 20 days prior to the admission.

Laboratory findings

White blood cell count: 5.10×10^9/L, lymphocyte count: 0.87×10^9/L, lymphocyte percentage: 17.1%, C-reactive protein: 74.2 mg/L.

Radiological findings

The radiological findings of bed-side DR during her hospitalization are as follows: Figs. 4.30–4.35.

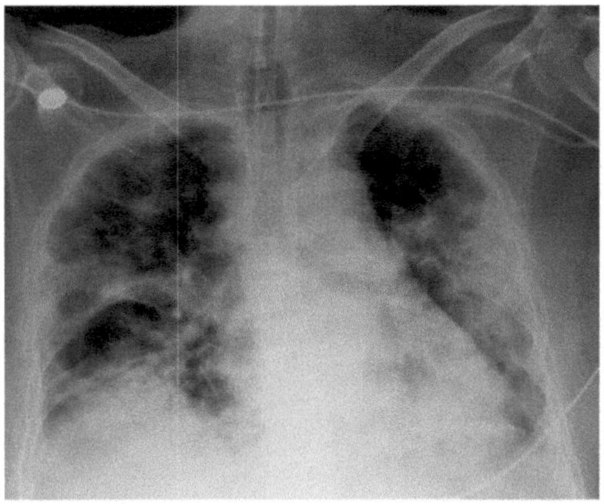

Figure 4.30 Bed-side DR on the day of admission. There are multiple patchy shadows in the bilateral upper lung fields.

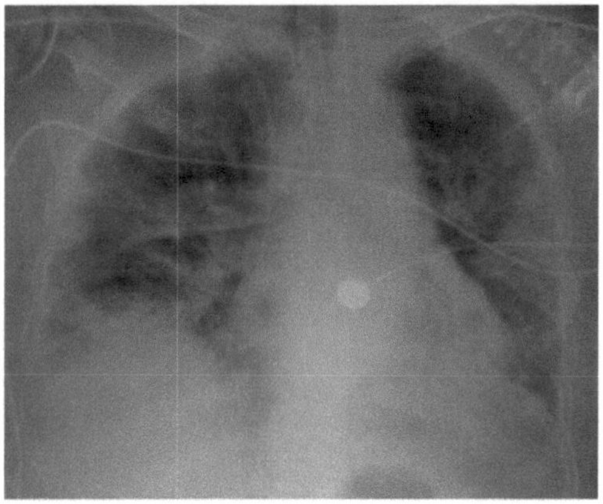

Figure 4.31 Bed-side DR 5 days after admission. There are multiple patchy shadows in the bilateral upper lung fields, while the lesions in the left upper lung show slight absorption.

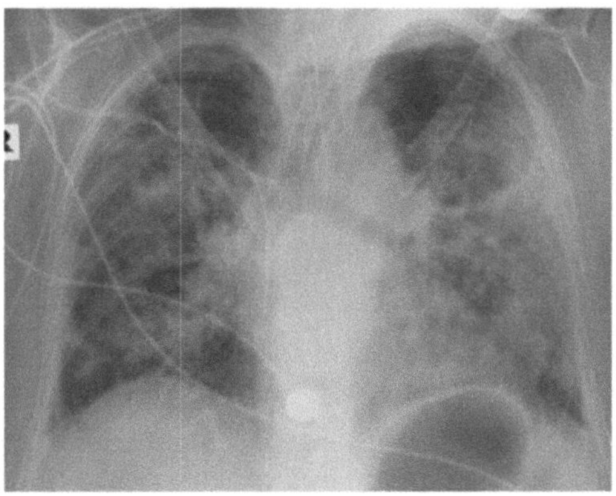

Figure 4.32 Bed-side DR 9 days after admission. The transparency of both lungs has decreased, with part of them showing white lung changes and enlargement of lesions.

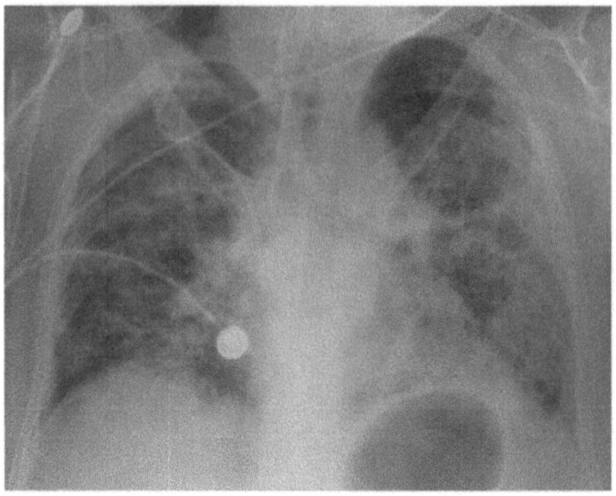

Figure 4.33 Bed-side DR 17 days after admission.

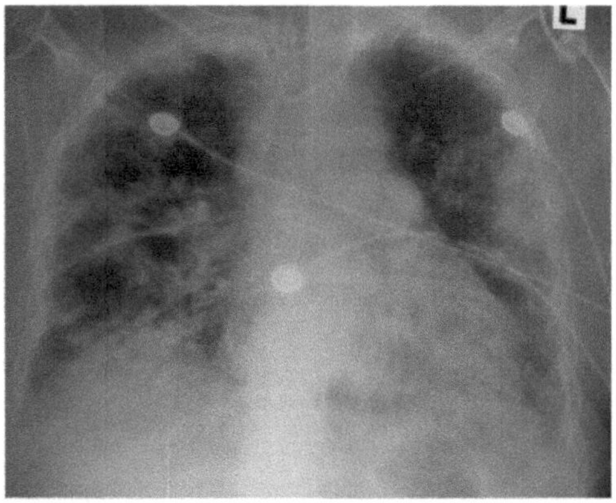

Figure 4.34 Bed-side DR 12 days after admission.

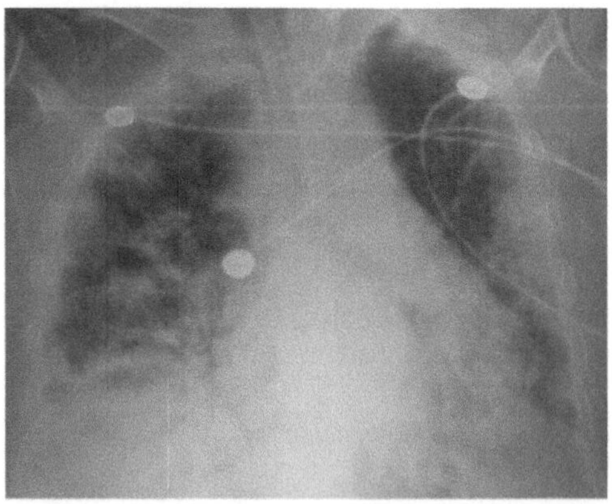

Figure 4.35 Bed-side DR 21 days after admission. The lesions in both lungs show a gradual reduction, with some of the patchy shadows turning into consolidations.

Case 59

Medical history

The patient is a 43-year-old man who documented a history of having had come in contact with people who were returned from the epidemic area. He reported fever 20 days prior to the admission date, with a maximum temperature of 38.0°C, accompanied by cough, chest tightness, expectoration, nausea, and vomiting. The nucleic acid test was positive 19 days prior to the admission.

Laboratory findings

White blood cell count: $9.80 \times 10^9/L$, lymphocyte count: $0.64 \times 10^9/L$, lymphocyte percentage: 6.5%, C-reactive protein: 200 mg/L.

Radiological findings

The radiological findings of bed-side DR during his hospitalization are as follows:

There are multiple patchy shadows in bilateral lung fields, with the greatest involvement on the day of admission and where there is gradual absorption afterward. There is a round bubble in the right lung fields with a liquid gas level in it, and gradually another bubble with a liquid gas level emerges in the right lung fields. Figs. 4.36 and 4.37.

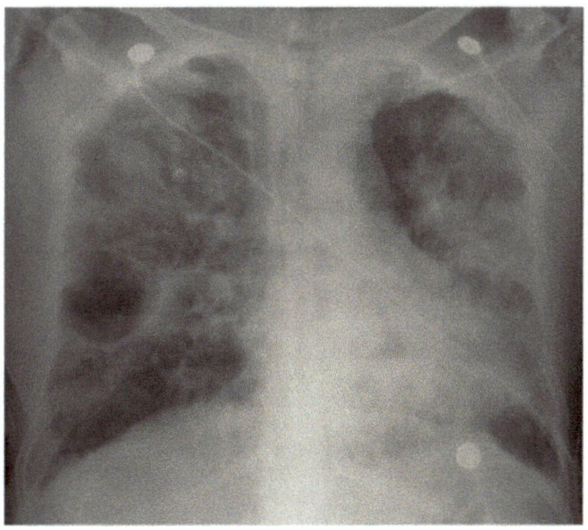

Figure 4.36 Bed-side DR on the day of admission. Multiple obscure attenuations are noted in bilateral lung fields. A round bulla is noted in the right middle lung field.

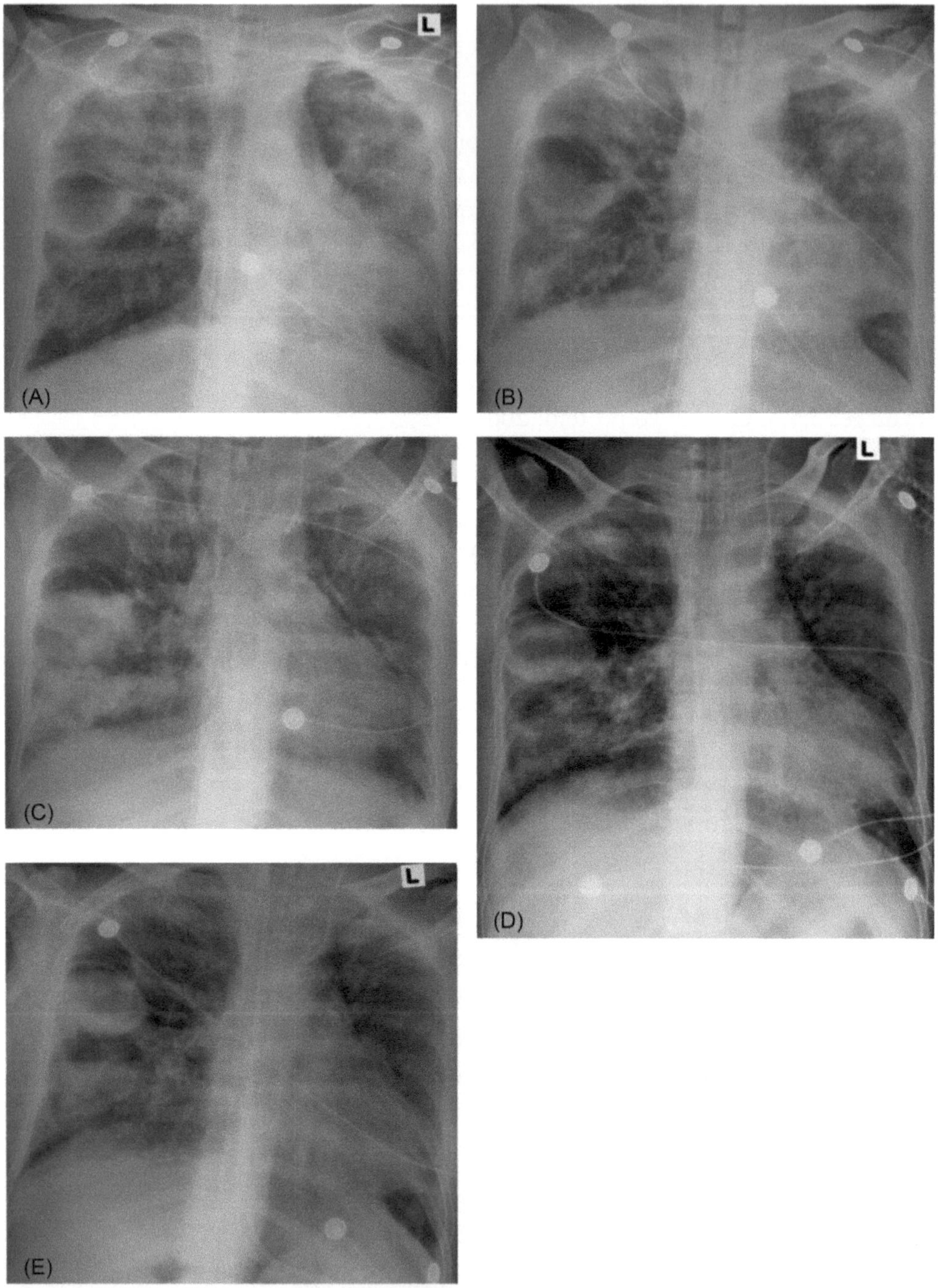

Figure 4.37 Bed-side DR (A) 5 days after admission, multiple obscure attenuations are noted in bilateral lung fields. A round bulla is noted in the right middle lung field. (B) 10 days after admission, fluid-level is seen in the bulla. (C) 14 days after admission, the lesions show absorption with previous image. (D) 18 days after admission, the lesions show continuous absorption, while the bulla is filled with high attenuation; and (E) 23 days after admission, another bulla with fluid level emerge in the right lower lung field.

Case 60

Medical history

The patient is a 54-year-old man who documented a history of contact with people diagnosed with COVID-19. He reported fever 20 days prior to the admission date, with a maximum temperature of 37.8°C, accompanied by cough and fatigue. The throat swab nucleic acid test was positive 9 days prior to the admission.

Laboratory findings

White blood cell count: $17.30 \times 10^9/L$, lymphocyte count: $1.52 \times 10^9/L$, lymphocyte percentage: 8.8%, C-reactive protein: 172.0 mg/L.

Radiological findings

The radiological findings of chest CT scans during his hospitalization are as follows: Figs. 4.38–4.41.

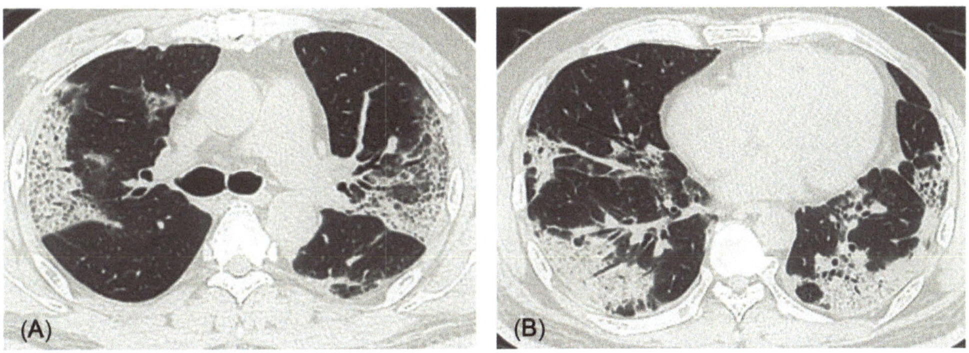

Figure 4.38 Chest computed tomography images 2 days after admission. (A) Multiple ground-glass opacities, consolidations and bands are noted in the outer zone of the upper lobes of both lungs, with bronchiectasis and crazypaving sign noted in the lesions. (B) Multiple ground-glass opacities and consolidations are noted in the bilateral lungs.

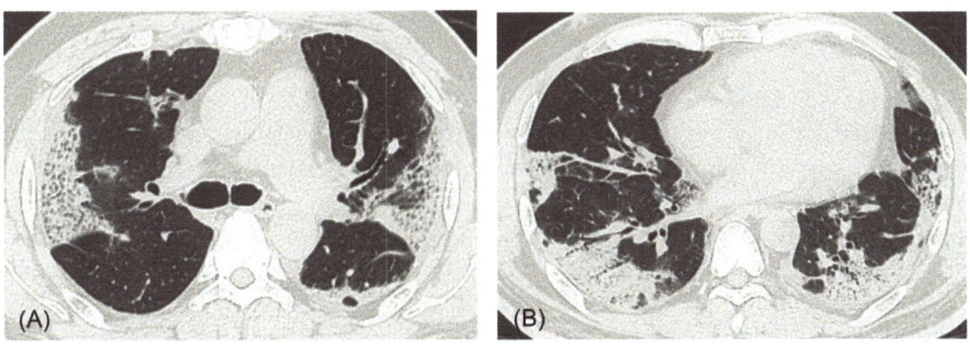

Figure 4.39 Chest computed tomography images 6 days after admission. (A) There is no obvious change in the lesions compared with those of prior images. (B) No obvious change is observed in bilateral lungs.

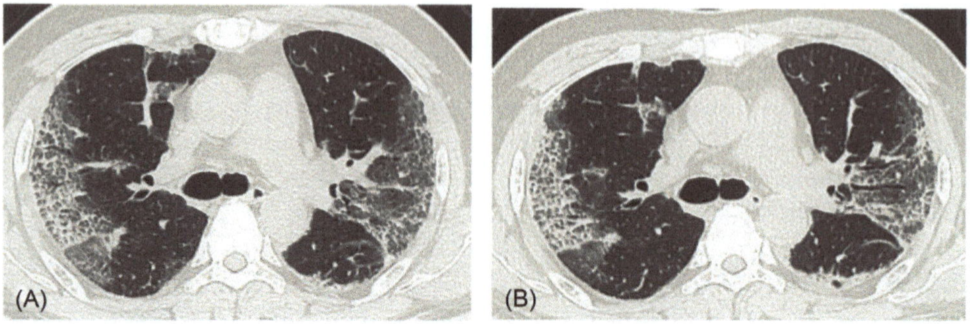

Figure 4.40 Chest computed tomography images 9 days after admission. (A) The lesions in both lungs show remarkable absorption, with the reduction of attenuation of most consolidations. The bronchiectasis is still there. (B) The lesions in both lungs show remarkable absorption, with most consolidations turning into ground-glass opacities.

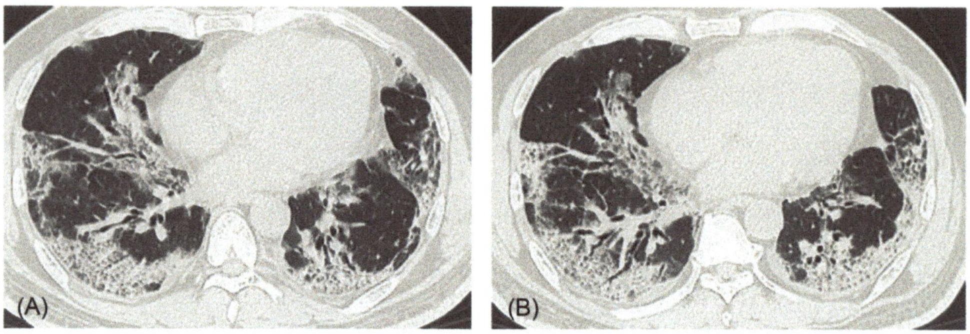

Figure 4.41 Chest computed tomography images 11 days after admission. (A) New emerging ground-glass opacities are noted in the right middle lobe. (B) The lesions in bilateral lower lobes show continuing absorption.

Case 61

Medical history

The patient is a 79-year-old woman who documented a history of contact with people diagnosed with COVID-19. She had diarrhea with tenesmus and abdominal pain 16 days prior to the admission. The above symptoms worsened 11 days prior to the admission date when she began purging with yellow watery stool, accompanied by a feeling of chest tightness, shortness of breath, nausea, and vomiting. The throat swab nucleic acid test was positive.

Laboratory findings

White blood cell count: $8.60 \times 10^9/L$, lymphocyte count: $1.91 \times 10^9/L$, lymphocyte percentage: 22.2%, C-reactive protein: 67.7 mg/L.

Radiological findings

The radiological findings of chest CT scans during her hospitalization are as follows: Figs. 4.42–4.44.

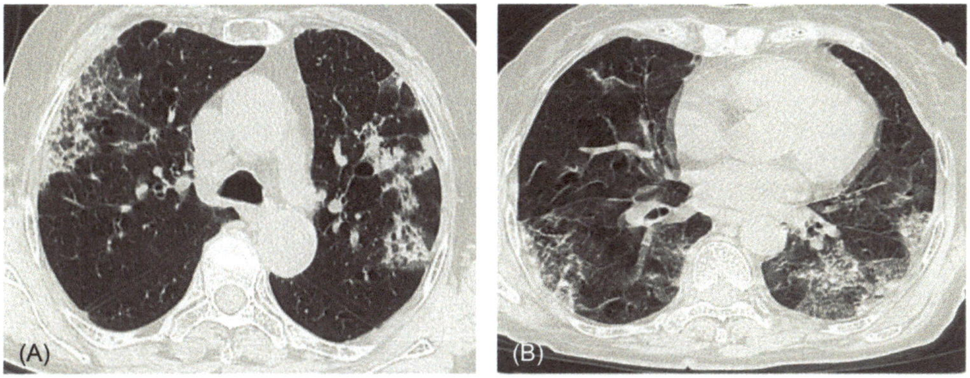

Figure 4.42 Chest computed tomography images 2 days after admission. (A) Multiple ground-glass opacities, consolidations and bands are noted in the upper lobes of both lungs, with bronchiolectasis seen inside. (B) Multiple ground-glass opacities and consolidations are noted in the lower lobes of both lungs (poor breath holding ability of the patient).

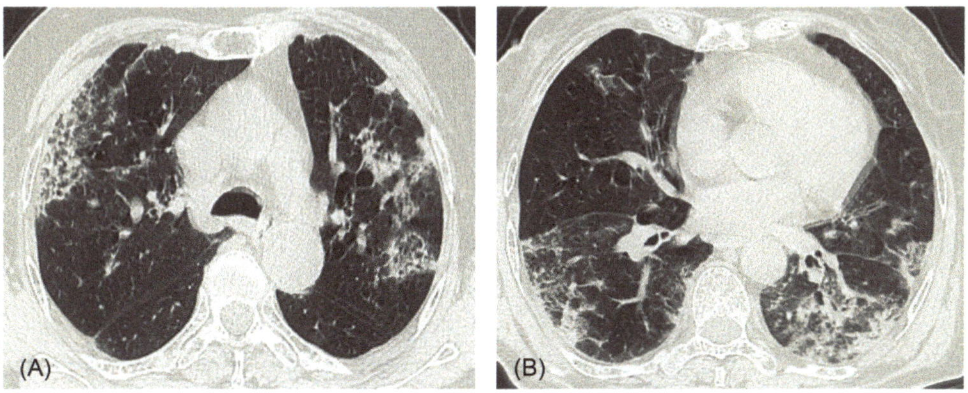

Figure 4.43 Chest computed tomography images 5 days after admission. (A) The lesions show similarity in area compared with those of prior images, whereas the attenuation of some consolidations shows a decrease. (B) Multiple ground-glass opacities, consolidation and streak shadows in the lower lobes of both lungs, with similar area of lesions and decreased lesion attenuation compared with previous images.

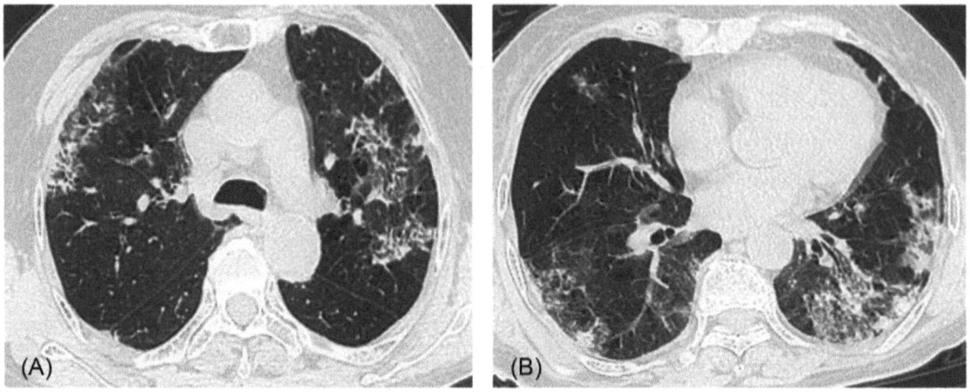

Figure 4.44 Chest computed tomography images 8 days after admission. (A) Multiple ground-glass opacities, consolidations and bands in the upper lobes of both lungs are noted, with bronchiolectasis seen inside. Compared to previous images, the lesions show continuous absorption. (B) The lesions in bilateral lower lobes show continuous absorption.

CHAPTER 5

Multimodality medical imaging in the diagnosis and evaluation of COVID-19

As the global COVID-19 pandemic progresses, imaging plays an important role in the diagnosis of COVID-19 patients. So far, multimodality medical imaging, including X-ray, computed tomography (CT), magnetic resonance imaging (MRI), positron emission tomography (PET/CT), and PET/magnetic resonance (MR) have been used in the diagnosis and evaluation of COVID-19.

X-ray is one of the first medical imaging methods to be used in clinical practice. Before the advent of CT, X-ray was the main diagnostic and evaluation tool for various human diseases, especially lung lesions. In the assessment of COVID-19, chest radiography (CXR) can not only show lesions initially but also appear to be widely available, inexpensive, fast, and easy to clean and maintain, helping to avoid cross-infection between patients [1]. So, globally, CXR is being used as a primary means of screening for COVID-19. Regarding CT, as a routine scanning technique for pneumonia, this method has been proved to be a useful tool for the diagnosis and identification of suspected patients with COVID-19 when still at an early stage. Compared with CXR, chest CT has higher sensitivity to detect COVID-19 because it allows the detection of some subtle lesions, including linear opacities, air bronchogram sign, bronchial wall thickening, interlobular septal thickening, and for reflecting interstitial lesions. A crazy-paving appearance may be investigated due to the lobular septal thickening and interstitial lesions. Chest CT may diagnose viral pneumonia earlier than RT-PCR, and some patients with negative RT-PCR results may present with positive chest CT findings for COVID-19. Furthermore, chest CT can evaluate the severity of the disease, which is conducive to the diagnosis and follow-up of COVID-19 pneumonia [2]. Although CT is the most sensitive way in radiologic evaluation of the lungs, CT exposes patients to ionizing radiation, while MRI does not. While MRI has some advantages over CT, such as high soft-tissue contrast, better tissue characteristics, and no ionizing radiation, still, at the same time, with the continuous development of imaging technology, the weakness of poor MRI signal and more artifacts in lung scanning has been gradually overcome with novel tools. For example, respiratory-gated ultrashort echo time MRI has the potential to play an important role similar to CT in the assessment of lung lesions. Therefore MRI is suitable for patients who need to avoid exposure to ionizing radiation, especially pregnant women, children,

Image Atlas of COVID-19
DOI: https://doi.org/10.1016/B978-0-443-18493-2.00005-X

and patients requiring dynamic monitoring. In addition to respiratory symptoms, COVID-19 can also affect other areas, such as the nervous system, where MRI has certain advantages in diagnosing these complications. For example, brain MRI usually shows acute and subacute infarcts. Other common findings include leukoencephalopathy and microhemorrhages leptomeningeal contrast enhancement, and cortical FLAIR signal abnormality [3]. ^{18}F-FDG PET/CT and PET/MR can provide metabolic information for lesions. Due to the accumulation of ^{18}F-FDG in activated inflammatory cells, ^{18}F-FDG PET/CT and PET/MR can be used for the diagnosis and monitoring of inflammatory diseases. It is reported that the typical findings of COVID-19 are increased uptake of ^{18}F-FDG in lesions formed from segmental ground-glass opacity as well as plaques and lymph nodes might be involved. In addition, ^{18}F-FDG PET/CT and PET/MR imaging can detect the increase of ^{18}F-FDG-uptake in other bodily organs of the COVID-19 patients, such as the liver and the spleen [4,5]. Thus, PET/CT and PET/MR imaging are useful in assessing the systemic involvement of COVID-19 patients, which seems to provide additional information about recovery and infection status in lungs and extrapulmonary organs.

X-ray, CT, MRI, PET/CT, and PET/MR can all contribute to the evaluation of COVID-19. In most cases, X-ray and CT are still the main tools for the diagnosis and evaluation of COVID-19. However, it may be difficult for a single imaging method to fully reflect the characteristics of the disease; therefore the application of various imaging modalities for a comprehensive assessment of the condition will provide maximum benefit to patients when needed.

References

[1] Balbi M, Caroli A, Corsi A, Milanese G, Surace A, Di Marco F, et al. Chest X-ray for predicting mortality and the need for ventilatory support in COVID-19 patients presenting to the emergency department. Eur Radiol 2021;31(4):1999−2012.

[2] Li K, Wu J, Wu F, Guo D, Chen L, Fang Z, et al. The clinical and chest CT features associated with severe and critical COVID-19 pneumonia. Invest Radiol 2020;55(6):327−31.

[3] Gulko E, Oleksk ML, Gomes W, Ali S, Mehta H, Overby P, et al. MRI brain findings in 126 patients with COVID-19: initial observations from a descriptive literature review. AJNR Am J Neuroradiol 2020;41(12):2199−203.

[4] Li X, Wang Y, Bai Y, Xu J, Fu C, Kang Y, et al. PET/MR and PET/CT in a severe COVID-19 patient. Eur J Nucl Med Mol Imaging 2020;47(10):2478−9.

[5] Bai Y, Xu J, Chen L, Fu C, Kang Y, Zhang W, et al. Inflammatory response in lungs and extrapulmonary sites detected by [^{18}F] fluorodeoxyglucose PET/CT in convalescing COVID-19 patients tested negative for coronavirus. Eur J Nucl Med Mol Imaging 2021;48(8):2531−42.

Index

Note: Page numbers followed by "*f*" refer to figures.

T

U

V

W

X

Y

MIX
Papier | Fördert gute Waldnutzung
FSC® C083411

Kolibri 360 GmbH
Ferdinand-Jühlke-Straße 7
99095 Erfurt, Deutschland
produktsicherheit@kolibri360.de